THE
VIOLINIST'S
THUMB

www.transworldbooks.co.uk

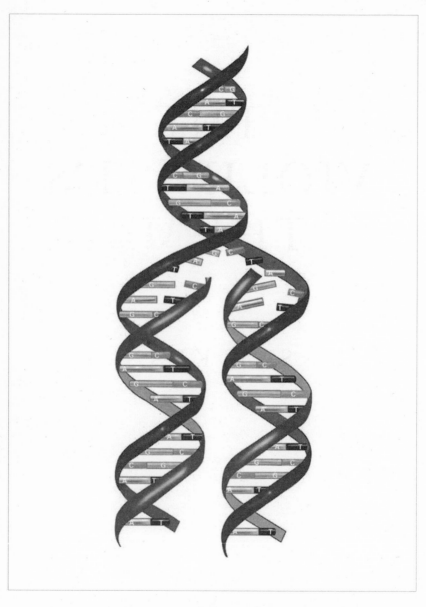

Life, therefore, may be considered a DNA chain reaction.

— MAXIM D. FRANK-KAMENETSKII,
UNRAVELING DNA

THE
VIOLINIST'S
THUMB

SAM KEAN

Doubleday

LONDON · TORONTO · SYDNEY · AUCKLAND · JOHANNESBURG

TRANSWORLD PUBLISHERS
61–63 Uxbridge Road, London W5 5SA
A Random House Group Company
www.transworldbooks.co.uk

First published in Great Britain
in 2012 by Doubleday
an imprint of Transworld Publishers

Frontispiece illustration by Mariana Ruiz and László Németh.

A CIP catalogue record for this book
is available from the British Library.

ISBN 9780857520289 (cased)
9780857520296 (tpb)

Addresses for Random House Group Ltd companies outside the UK
can be found at: www.randomhouse.co.uk
The Random House Group Ltd Reg. No. 954009

The Random House Group Limited supports the Forest Stewardship Council (FSC®), the
leading international forest-certification organization. Our books carrying the FSC label
are printed on FSC®-certified paper. FSC is the only forest-certification scheme endorsed
bythe leading environmental organizations, including Greenpeace. Our paper procurement
policy can be found at www.randomhouse.co.uk/environment

Typeset in Janson
Printed and bound in Great Britain by
CPI Group (UK) Ltd, Croydon, CR0 4YY

2 4 6 8 10 9 7 5 3 1

Acrostic: *n., an incognito message formed by stringing together the initial letters of lines or paragraphs or other units of composition in a work.*

N.B.: I've hidden a DNA-related acrostic in *The Violinist's Thumb* — a genetic "Easter egg," if you will. If you decode this message, e-mail me through my website (http://samkean.com /contact). Or if you can't figure it out, e-mail me anyway and I'll reveal the answer.

Also by Sam Kean

The Disappearing Spoon

For more information on Sam Kean and his books,
see his website at www.samkean.com

CONTENTS

PART III
GENES AND GENIUSES:
HOW HUMANS BECAME ALL TOO HUMAN

PART IV
THE ORACLE OF DNA:
GENETICS IN THE PAST, PRESENT, AND FUTURE

CONTENTS

THE
VIOLINIST'S
THUMB

Introduction

This might as well come out up front, first paragraph. This is a book about DNA—about digging up stories buried in your DNA for thousands, even millions of years, and using DNA to solve mysteries about human beings whose solutions once seemed lost forever. And yes, I'm writing this book despite the fact that my father's name is Gene. As is my mother's name. Gene and Jean. Gene and Jean Kean. Beyond being singsong absurd, my parents' names led to a lot of playground jabs over the years: my every fault and foible was traced to "my genes," and when I did something idiotic, people smirked that "my genes made me do it." That my parents' passing on their genes necessarily involved sex didn't help. The taunts were doubly barbed, utterly unanswerable.

Bottom line is, I dreaded learning about DNA and genes in science classes growing up because I knew some witticism would be coming within about two seconds of the teacher turning her back. And if it wasn't coming, some wiseacre was *thinking* it. Some of that Pavlovian trepidation always stayed with me, even when (or especially when) I began to grasp how potent a substance DNA is. I got over the gibes by high school, but the word *gene* still evoked a lot of simultaneous responses, some agreeable, some not.

On the one hand, DNA excites me. There's no bolder topic in science than genetics, no field that promises to push science forward to the same degree. I don't mean just the common (and commonly overblown) promises of medical cures, either. DNA has revitalized every field in biology and remade the very study of human beings. At the same time, whenever someone starts digging into our basic human biology, we resist the intrusion— we don't want to be reduced to mere DNA. And when someone talks about tinkering with that basic biology, it can be downright frightening.

More ambiguously, DNA offers a powerful tool for rooting through our past: biology has become history by other means. Even in the past decade or so, genetics has opened up a whole Bible's worth of stories whose plotlines we assumed had vanished— either too much time had lapsed, or too little fossil or anthropological evidence remained to piece together a coherent narrative. It turns out we were carrying those stories with us the entire time, trillions of faithfully recorded texts that the little monks in our cells transcribed every hour of every day of our DNA dark age, waiting for us to get up to speed on the language. These stories include the grand sagas of where we came from and how we evolved from primordial muck into the most dominant species the planet has known. But the stories come home in surprisingly individual ways, too.

If I could have had one mulligan in school (besides a chance to make up safer names for my parents), I'd have picked a different instrument to play in band. It wasn't because I was the only boy clarinetist in the fourth, fifth, sixth, seventh, eighth, and ninth grades (or not only because of that). It was more because I felt so clumsy working all the valves and levers and blowholes on the clarinet. Nothing to do with a lack of practice, surely. I blamed the deficit on my double-jointed fingers and splayed hitchhiker thumbs. Playing the clarinet wound my fingers into such awkward braids that I constantly felt a need to crack my knuckles, and they'd throb a little. Every blue moon one thumb would even get stuck in place, frozen in extension, and I had to work the joint free with my other hand. My fingers just didn't do what the better girl clarinetists' could. My problems were inherited, I told myself, a legacy of my parents' gene stock.

After quitting band, I had no reason to reflect on my theory about manual dexterity and musical ability until a decade later, when I learned the story of violinist Niccolò Paganini, a man so gifted he had to shake off rumors his whole life that he'd sold his soul to Satan for his talent. (His hometown church even refused to bury his body for decades after his death.) It turns out Paganini had made a pact with a subtler master, his DNA. Paganini almost certainly had a genetic disorder that gave him freakishly flexible fingers. His connective tissues were so rubbery that he could pull his pinky out sideways to form a right angle to the rest of his hand. (Try this.) He could also stretch his hands abnormally wide, an incomparable advantage when playing the violin. My simple hypothesis about people "being born" to play (or not play) certain instruments seemed justified. I should have quit when ahead. I kept investigating and found out that Paganini's syndrome probably caused serious health problems, as joint pain, poor vision, weakness of breath, and fatigue dogged the violinist his whole life. I whimpered about stiff knuckles during early a.m. marching-band practice, but Paganini frequently had to cancel shows at the height of his career and couldn't perform in public during the last years of his life. In Paganini, a passion for music had united with a body perfectly tuned to take advantage of its flaws, possibly the greatest fate a human could hope for. Those flaws then hastened his death. Paganini may not have chosen his pact with his genes, but he was in one, like all of us, and the pact both made and unmade him.

DNA wasn't done telling its stories to me. Some scientists have retroactively diagnosed Charles Darwin, Abraham Lincoln, and Egyptian pharaohs with genetic disorders. Other scientists have plumbed DNA itself to articulate its deep linguistic properties and surprising mathematical beauty. In fact, just as I had crisscrossed from band to biology to history to math to social studies in high school, so stories about DNA began popping up in all sorts of contexts, linking all sorts of disparate subjects. DNA informed stories about people surviving nuclear bombs, and stories about the untimely ends of explorers in the Arctic. Stories about the near extinction of the human species, or pregnant mothers giving cancer to their unborn children. Stories where, as with Paganini, science illuminates art, and even stories where — as with scholars tracing genetic defects through portraiture — art illuminates science.

One fact you learn in biology class but don't appreciate at first is the sheer length of the DNA molecule. Despite being packed into a tiny closet in our already tiny cells, DNA can unravel to extraordinary distances. There's enough DNA in some plant cells to stretch three hundred feet; enough DNA in one human body to stretch roughly from Pluto to the sun and back; enough DNA on earth to stretch across the known universe many, many times. And the further I pursued the stories of DNA, the more I saw that its quality of stretching on and on — of unspooling farther and farther out, and even back, back through time — was intrinsic to DNA. Every human activity leaves a forensic trace in our DNA, and whether that DNA records stories about music or sports or Machiavellian microbes, those tales tell, collectively, a larger and more intricate tale of the rise of human beings on Earth: why we're one of nature's most absurd creatures, as well as its crowning glory.

Underlying my excitement, though, is the other side of genes: the trepidation. While researching this book, I submitted my DNA to a genetic testing service, and despite the price tag

($414), I did so in a frivolous state of mind. I knew personal genomic testing has serious shortcomings, and even when the science is solid, it's often not that helpful. I might learn from my DNA that I have green eyes, but then again I do own a mirror. I might learn I don't metabolize caffeine well, but I've had plenty of jittery nights after a late Coke. Besides, it was hard to take the DNA-submission process seriously. A plastic vial with a candy-corn orange lid arrived in the mail, and the instructions told me to massage my cheeks with my knuckles to work some cells loose inside my mouth. I then hocked into the tube repeatedly until I filled it two-thirds full of saliva. That took ten minutes, since the instructions said in all seriousness that it couldn't be just any saliva. It had to be the good, thick, syrupy stuff; as with a draft beer, there shouldn't be much foam. The next day I mailed the genetic spittoon off, hoping for a nice surprise about my ancestry. I didn't engage in any sober reflection until I went to register my test online and read the instructions about redacting sensitive or scary information. If your family has a history of breast cancer or Alzheimer's or other diseases—or if the mere thought of having them frightens you—the testing service lets you block that information. You can tick a box and keep it secret from even yourself. What caught me short was the box for Parkinson's disease. One of the earliest memories I have, and easily the worst of those early memories, is wandering down the hallway of my grandma's house and poking my head into the room where my grandpa, laid low by Parkinson's, lived out his days.

When he was growing up, people always told my father how much he looked like my grandpa—and I got similar comments about looking like my old man. So when I wandered into that room off the hallway and saw a white-haired version of my father propped in a bed with a metal safety rail, I saw myself by extension. I remember lots of white—the walls, the carpet, the sheets, the open-backed smock he wore. I remember him pitched forward to the point of almost tipping over, his smock loose and a fringe of white hair hanging straight down.

I'm not sure whether he saw me, but when I hesitated on the threshold, he moaned and began trembling, which made his voice quake. My grandpa was lucky in some ways; my grandma, a nurse, took care of him at home, and his children visited regularly. But he'd regressed mentally and physically. I remember most of all the thick, syrupy string of saliva pendulous on his chin, full of DNA. I was five or so, too young to understand. I'm still ashamed that I ran.

Now, strangers—and worse, my own self—could peek at whether the string of self-replicating molecules that might have triggered Parkinson's in my grandfather was lurking in my cells, too. There was a good chance not. My grandpa's genes had been diluted by my grandma's genes in Gene, whose genes had in turn been diluted in me by Jean's. But the chance was certainly real. I could face any of the cancers or other degenerative diseases I might be susceptible to. Not Parkinson's. I blacked the data out.

Personal stories like that are as much a part of genetics as all the exciting history—perhaps more so, since all of us have at least one of these stories buried inside us. That's why this book, beyond relating all the historical tales, builds on those tales and links them to work being done on DNA today, and work likely to be done tomorrow. This genetics research and the changes it will bring have been compared to a shifting ocean tide, huge and inevitable. But its consequences will arrive at the shore where we're standing not as a tsunami but as tiny waves. It's the individual waves we'll feel, one by one, as the tide crawls up the shore, no matter how far back we think we can stand.

Still, we can prepare ourselves for their arrival. As some scientists recognize, the story of DNA has effectively replaced the old college Western Civ class as the grand narrative of human existence. Understanding DNA can help us understand where we come from and how our bodies and minds work, and under-standing the limits of DNA also helps us understand how our bodies and minds don't work. To a similar degree, we'll have to prepare ourselves for whatever DNA says (and doesn't say) about intractable social problems like gender and race relations, or whether traits like aggression and intelligence are fixed or flexi-ble. We'll also have to decide whether to trust eager thinkers who, while acknowledging that we don't understand completely how DNA works, already talk about the opportunity, even the obligation, to improve on four billion years of biology. To this point of view, the most remarkable story about DNA is that our species survived long enough to (potentially) master it.

The history in this book is still being constructed, and I structured *The Violinist's Thumb* so that each chapter provides the answer to a single question. The overarching narrative starts in the remote microbial past, moves on to our animal ancestries, lingers over primates and hominid competitors like Neander-thals, and culminates with the emergence of modern, cultured human beings with flowery language and hypertrophied brains. But as the book advances toward the final section, the questions have not been fully resolved. Things remain uncertain—especially the question of how this grand human experiment of uprooting everything there is to know about our DNA will turn out.

PART I

A, C, G, T, and You

How to Read a Genetic Score

1

Genes, Freaks, DNA

How Do Living Things Pass Down Traits to Their Children?

hills and flames, frost and inferno, fire and ice. The two scientists who made the first great discoveries in genetics had a lot in common — not least the fact that both died obscure, mostly unmourned and happily forgotten by many. But whereas one's legacy perished in fire, the other's succumbed to ice.

The blaze came during the winter of 1884, at a monastery in what's now the Czech Republic. The friars spent a January day emptying out the office of their deceased abbot, Gregor Mendel, ruthlessly purging his files, consigning everything to a bonfire in the courtyard. Though a warm and capable man, late in life Mendel had become something of an embarrassment to the monastery, the cause for government inquiries, newspaper gossip, even a showdown with a local sheriff. (Mendel won.) No relatives came by to pick up Mendel's things, and the monks burned his papers for the same reason you'd cauterize a wound — to sterilize, and stanch embarrassment. No record survives of what they looked like, but among those documents were sheaves of papers, or perhaps a lab notebook with a plain cover, probably coated in dust from disuse. The yellowed pages would have been full of sketches of pea plants and tables of numbers (Mendel adored numbers), and they probably didn't kick up any more smoke and ash than other papers when incinerated. But the burning of those papers—burned on the exact spot where Mendel had kept his greenhouse years before—destroyed the only original record of the discovery of the gene.

The chills came during that same winter of 1884—as they had for many winters before, and would for too few winters after. Johannes Friedrich Miescher, a middling professor of physiology in Switzerland, was studying salmon, and among his other projects he was indulging a long-standing obsession with a substance—a cottony gray paste—he'd extracted from salmon sperm years before. To keep the delicate sperm from perishing in the open air, Miescher had to throw the windows open to the cold and refrigerate his lab the old-fashioned way, exposing himself day in and day out to the Swiss winter. Getting any work done required superhuman focus, and that was the one asset even people who thought little of Miescher would admit he had. (Earlier in his career, friends had to drag him from his lab bench one afternoon to attend his wedding; the ceremony had slipped his mind.) Despite being so driven, Miescher had pathetically little to show for it—his lifetime scientific output was meager. Still, he kept the windows open and kept shivering year after year, though he knew it was slowly killing him. And he still never got to the bottom of that milky gray substance, DNA.

DNA and genes, genes and DNA. Nowadays the words have become synonymous. The mind rushes to link them, like Gilbert and Sullivan or Watson and Crick. So it seems fitting that Miescher and Mendel discovered DNA and genes almost simultaneously in the 1860s, two monastic men just four hundred miles apart in the German-speaking span of middle Europe. It seems more than fitting; it seems fated.

But to understand what DNA and genes really are, we have to decouple the two words. They're not identical and never have been. DNA is a *thing*—a chemical that sticks to your fingers. Genes have a physical nature, too; in fact, they're made of long stretches of DNA. But in some ways genes are better viewed as conceptual, not material. A gene is really information—more like a story, with DNA as the language the story is written in. DNA and genes combine to form larger structures called chromosomes, DNA-rich volumes that house most of the genes in living things. Chromosomes in turn reside in the cell nucleus, a library with instructions that run our entire bodies.

All these structures play important roles in genetics and heredity, but despite the near-simultaneous discovery of each in the 1800s, no one connected DNA and genes for almost a century, and both discoverers died uncelebrated. How biologists finally yoked genes and DNA together is the first epic story in the science of inheritance, and even today, efforts to refine the relationship between genes and DNA drive genetics forward.

Mendel and Miescher began their work at a time when folk theories—some uproarious or bizarre, some quite ingenious, in their way—dominated most people's thinking about heredity, and for centuries these folk theories had colored their views about why we inherit different traits.

Everyone knew on some level of course that children resemble parents. Red hair, baldness, lunacy, receding chins, even extra thumbs, could all be traced up and down a genealogical tree. And fairy tales—those codifiers of the collective unconscious—often turned on some wretch being a "true" prince(ss) with a royal bloodline, a biological core that neither rags nor an amphibian frame could sully.

That's mostly common sense. But the mechanism of heredity—how exactly traits got passed from generation to generation—baffled even the most intelligent thinkers, and the vagaries of this process led to many of the wilder theories that circulated before and even during the 1800s. One ubiquitous folk theory, "maternal impressions," held that if a pregnant woman saw something ghoulish or suffered intense emotions, the experience would scar her child. One woman who never satisfied an intense prenatal craving for strawberries gave birth to a baby covered with red, strawberry-shaped splotches. The same could happen with bacon. Another woman bashed her head on a sack of coal, and her child had half, but only half, a head of black hair. More direly, doctors in the 1600s reported that a woman in Naples, after being startled by sea monsters, bore a son covered in scales, who ate fish exclusively and gave off fishy odors. Bishops told cautionary tales of a woman who seduced her actor husband backstage in full costume. He was playing Mephistopheles; they had a child with hooves and horns. A beggar with one arm spooked a woman into having a one-armed child. Pregnant women who pulled off crowded streets to pee in churchyards invariably produced bed wetters. Carrying fireplace logs about in your apron, next to the bulging tummy, would produce a grotesquely well-hung lad. About the only recorded happy case of maternal impressions involved a patriotic woman in Paris in the 1790s whose son had a birthmark on his chest shaped like a Phrygian cap—those elfish hats with a flop of material on top. Phrygian caps were symbols of freedom to the new French republic, and the delighted government awarded her a lifetime pension.

Much of this folklore intersected with religious belief, and people naturally interpreted serious birth defects—cyclopean eyes, external hearts, full coats of body hair—as back-of-the-Bible warnings about sin, wrath, and divine justice. One example from the 1680s involved a cruel bailiff in Scotland named Bell, who arrested two female religious dissenters, lashed them to poles near the shore, and let the tide swallow them. Bell added insult by taunting the women, then drowned the younger, more stubborn one with his own hands. Later, when asked about the murders, Bell always laughed, joking that the women must be having a high time now, scuttling around among the crabs. The joke was on Bell: after he married, his children were born with a severe defect that twisted their forearms into two awful pincers. These crab claws proved highly heritable to their children and grandchildren, too. It didn't take a biblical scholar to see that the iniquity of the father had been visited upon the children, unto the third and fourth generations. (And beyond: cases popped up in Scotland as late as 1900.)

If maternal impressions stressed environmental influences, other theories of inheritance had strong congenital flavors. One, preformationism, grew out of the medieval alchemists' quest to create a homunculus, a miniature, even microscopic, human being. Homunculi were the biological philosopher's stone, and creating one showed that an alchemist possessed the power of gods. (The process of creation was somewhat less dignified. One recipe called for fermenting sperm, horse dung, and urine in a pumpkin for six weeks.) By the late 1600s, some protoscientists had stolen the idea of the homunculus and were arguing that one must live inside each female egg cell. This neatly did away with the question of how living embryos arose from seemingly dead blobs of matter. Under preformationist theory, such spontaneous generation wasn't necessary: homuncular babies were indeed preformed and merely needed a trigger, like sperm, to grow. This idea had only one problem: as critics pointed out, it introduced an infinite regress, since a woman necessarily had to have all her future children, as well as their children, and *their* children, stuffed inside her, like Russian *matryoshka* nesting dolls. Indeed, adherents of "ovism" could only deduce that God had crammed the entire human race into Eve's ovaries on day one. (Or rather, day six of Genesis.) "Spermists" had it even worse— Adam must have had humanity entire sardined into his even tinier sperms. Yet after the first microscopes appeared, a few spermists tricked themselves into seeing tiny humans bobbing around in puddles of semen. Both ovism and spermism gained credence in part because they explained original sin: we all resided inside Adam or Eve during their banishment from Eden and therefore all share the taint. But spermism also introduced theological quandaries—for what happened to the endless number of unbaptized souls that perished every time a man ejaculated?

However poetic or deliciously bawdy these theories were, biologists in Miescher's day scoffed at them as old wives' tales. These men wanted to banish wild anecdotes and vague "life forces" from science and ground all heredity and development in chemistry instead.

Miescher hadn't originally planned to join this movement to demystify life. As a young man he had trained to practice the family trade, medicine, in his native Switzerland. But a boyhood typhoid infection had left him hard of hearing and unable to use a stethoscope or hear an invalid's bedside bellyaching. Miescher's father, a prominent gynecologist, suggested a career in research instead. So in 1868 the young Miescher moved into a lab run by the biochemist Felix Hoppe-Seyler, in Tübingen, Germany. Though headquartered in an impressive medieval castle, Hoppe-Seyler's lab occupied the royal laundry room in the basement; he found Miescher space next door, in the old kitchen.

Hoppe-Seyler wanted to catalog the chemicals present in human blood cells. He had already investigated red blood cells, so he assigned white ones to Miescher—a fortuitous decision

Friedrich Miescher (*inset*) discovered DNA in this laboratory, a renovated kitchen in the basement of a castle in Tübingen, Germany. (University of Tübingen library)

for his new assistant, since white blood cells (unlike red ones) contain a tiny internal capsule called a nucleus. At the time, most scientists ignored the nucleus—it had no known function—and quite reasonably concentrated on the cytoplasm instead, the slurry that makes up most of a cell's volume. But the chance to analyze something unknown appealed to Miescher.

To study the nucleus, Miescher needed a steady supply of white blood cells, so he approached a local hospital. According to legend, the hospital catered to veterans who'd endured gruesome battlefield amputations and other mishaps. Regardless, the clinic did house many chronic patients, and each day a hospital orderly collected pus-soaked bandages and delivered the yellowed rags to Miescher. The pus often degraded into slime in the open air, and Miescher had to smell each suppurated-on cloth and throw out the putrid ones (most of them). But the remaining "fresh" pus was swimming with white blood cells.

Eager to impress—and, in truth, doubtful of his own talents—Miescher threw himself into studying the nucleus, as if sheer labor would make up for any shortcomings. A colleague later described him as "driven by a demon," and Miescher exposed himself daily to all manner of chemicals in his work. But without this focus, he probably wouldn't have discovered what he did, since the key substance inside the nucleus proved elusive. Miescher first washed his pus in warm alcohol, then acid extract from a pig's stomach, to dissolve away the cell membranes. This allowed him to isolate a gray paste. Assuming it was protein, he ran tests to identify it. But the paste resisted protein digestion and, unlike any known protein, wouldn't dissolve in salt water, boiling vinegar, or dilute hydrochloric acid. So he tried elementary analysis, charring it until it decomposed. He got the expected elements, carbon, hydrogen, oxygen, and nitrogen, but also discovered 3 percent phosphorus, an element proteins lack. Convinced he'd found something unique, he named the substance "nuclein"—what later scientists called deoxyribonucleic acid, or DNA.

Miescher polished off the work in a year, and in autumn 1869 stopped by the royal laundry to show Hoppe-Seyler. Far from rejoicing, the older scientist screwed up his brow and expressed his doubts that the nucleus contained any sort of special, nonproteinaceous substance. Miescher had made a mistake, surely. Miescher protested, but Hoppe-Seyler insisted on repeating the young man's experiments—step by step, bandage by bandage—

before allowing him to publish. Hoppe-Seyler's condescension couldn't have helped Miescher's confidence (he never worked so quickly again). And even after two years of labor vindicated Miescher, Hoppe-Seyler insisted on writing a patronizing editorial to accompany Miescher's paper, in which he backhandedly praised Miescher for "enhanc[ing] our understanding...of pus." Nevertheless Miescher did get credit, in 1871, for discovering DNA.

Some parallel discoveries quickly illuminated more about Miescher's molecule. Most important, a German protégé of Hoppe-Seyler's determined that nuclein contained multiple types of smaller constituent molecules. These included phosphates and sugars (the eponymous "deoxyribose" sugars), as well as four ringed chemicals now called nucleic "bases"—adenine, cytosine, guanine, and thymine. Still, no one knew how these parts fit together, and this jumble made DNA seem strangely heterogeneous and incomprehensible.

(Scientists now know how all these parts contribute to DNA. The molecule forms a double helix, which looks like a ladder twisted into a corkscrew. The supports of the ladder are strands made of alternating phosphates and sugars. The ladder's rungs— the most important part—are each made of two nucleic bases, and these bases pair up in specific ways: adenine, A, always bonds with thymine, T; cytosine, C, always bonds with guanine, G. [To remember this, notice that the curvaceous letters C and G pair-bond, as do angular A and T.])

Meanwhile DNA's reputation was bolstered by other discoveries. Scientists in the later 1800s determined that whenever cells divide in two, they carefully divvy up their chromosomes. This hinted that chromosomes were important for something, because otherwise cells wouldn't bother. Another group of scientists determined that chromosomes are passed whole and intact from parent to child. Yet another German chemist then discovered that chromosomes were mostly made up of none other than DNA. From this constellation of findings—it took a little imagination to sketch in the lines and see a bigger picture—a small number of scientists realized that DNA might play a direct role in heredity. Nuclein was intriguing people.

Miescher lucked out, frankly, when nuclein became a respectable object of inquiry; his career had stalled otherwise. After his stint in Tübingen, he moved home to Basel, but his new institute refused him his own lab—he got one corner in a common room and had to carry out chemical analyses in an old hallway. (The castle kitchen was looking pretty good suddenly.) His new job also required teaching. Miescher had an aloof, even frosty demeanor—he was someone never at ease around people—and although he labored over lectures, he proved a pedagogical disaster: students remember him as "insecure, restless...myopic... difficult to understand, [and] fidgety." We like to think of scientific heroes as electric personalities, but Miescher lacked even rudimentary charisma.

Given his atrocious teaching, which further eroded his self-esteem, Miescher rededicated himself to research. Upholding what one observer called his "fetish of examining objectionable fluids," Miescher transferred his DNA allegiance from pus to semen. The sperm in semen were basically nuclein-tipped missiles and provided loads of DNA without much extraneous cytoplasm. Miescher also had a convenient source of sperm in the hordes of salmon that clogged the river Rhine near his university every autumn and winter. During spawning season, salmon testes grow like tumors, swelling twenty times larger than normal and often topping a pound each. To collect salmon, Miescher could practically dangle a fishing line from his office window, and by squeezing their "ripe" testes through cheesecloth, he isolated millions of bewildered little swimmers. The downside was that salmon sperm deteriorates at anything close to comfortable temperatures. So Miescher had to arrive at his bench in the chilly early hours before dawn, prop the windows open, and drop the temperature to around 35°F before working. And because of a stingy budget, when his laboratory glassware broke, he sometimes had to pilfer his ever-loving wife's fine china to finish experiments.

From this work, as well as his colleagues' work with other cells, Miescher concluded that all cell nuclei contain DNA. In fact he proposed redefining cell nuclei—which come in a variety of sizes and shapes—strictly as containers for DNA. Though he wasn't greedy about his reputation, this might have been a last stab at glory for Miescher. DNA might still have turned out to be relatively unimportant, and in that case, he would have at least figured out what the mysterious nucleus did. But it wasn't to be. Though we now know Miescher was largely right in defining the nucleus, other scientists balked at his admittedly premature suggestion; there just wasn't enough proof. And even if they bought that, they wouldn't grant Miescher's next, more self-serving claim: that DNA influenced heredity. It didn't help that Miescher had no idea how DNA did so. Like many scientists then, he doubted that sperm injected anything into eggs, partly because he assumed (echoes of the homunculus here) that eggs already contained the full complement of parts needed for life. Rather, he imagined that sperm nuclein acted as a sort of chemical defibrillator and jump-started eggs. Unfortunately Miescher had little time to explore or defend such ideas. He still had to lecture, and the Swiss government piled "thankless and tedious" tasks onto him, like preparing reports on nutrition in prisons and elementary schools. The years of working through Swiss winters with the windows open also did a number on his health, and he contracted tuberculosis. He ended up giving up DNA work altogether.

Meanwhile other scientists' doubts about DNA began to

solidify, in their minds, into hard opposition. Most damning, scientists discovered that there was more to chromosomes than phosphate-sugar backbones and A-C-G-T bases. Chromosomes also contained protein nuggets, which seemed more likely candidates to explain chemical heredity. That's because proteins were composed of twenty different subunits (called amino acids). Each of these subunits could serve as one "letter" for writing chemical instructions, and there seemed to be enough variety among these letters to explain the dazzling diversity of life itself. The A, C, G, and T of DNA seemed dull and simplistic in comparison, a four-letter pidgin alphabet with limited expressive power. As a result, most scientists decided that DNA stored phosphorus for cells, nothing more.

Sadly, even Miescher came to doubt that DNA contained enough alphabetical variety. He too began tinkering with protein inheritance, and developed a theory where proteins encoded information by sticking out molecular arms and branches at different angles—a kind of chemical semaphore. It still wasn't clear how sperm passed this information to eggs, though, and Miescher's confusion deepened. He turned back to DNA late in life and argued that it might assist with heredity still. But progress proved slow, partly because he had to spend more and more time in tuberculosis sanitariums in the Alps. Before he got to the bottom of anything, he contracted pneumonia in 1895, and succumbed soon after.

Later work continued to undermine Miescher by reinforcing the belief that even if chromosomes control inheritance, the proteins in chromosomes, not the DNA, contained the actual information. After Miescher's death, his uncle, a fellow scientist, gathered Miescher's correspondence and papers into a "collected works," like some belle-lettrist. The uncle prefaced the book by claiming that "Miescher and his work will not diminish; on the contrary, it will grow and his discoveries and thoughts will be seeds for a fruitful future." Kind words, but it must have seemed a fond hope: Miescher's obituaries barely mentioned his work on nuclein; and DNA, like Miescher himself, seemed decidedly minor.

At least Miescher died known, where he was known, for science. Gregor Mendel made a name for himself during his lifetime only through scandal.

By his own admission, Mendel became an Augustinian friar not because of any pious impulse but because his order would pay his bills, including college tuition. The son of peasants, Mendel had been able to afford his elementary school only because his uncle had founded it; he attended high school only after his sister sacrificed part of her dowry. But with the church footing the bill, Mendel attended the University of Vienna and studied science, learning experimental design from Christian Doppler himself, of the eponymous effect. (Though only after Doppler rejected Mendel's initial application, perhaps because of Mendel's habit of having nervous breakdowns during tests.)

The abbot at St. Thomas, Mendel's monastery, encouraged Mendel's interest in science and statistics, partly for mercenary reasons: the abbot thought scientific farming could produce better sheep, fruit trees, and grapevines and help the monastery crawl out of debt. But Mendel had time to explore other interests, too, and over the years he charted sunspots, tracked tornadoes, kept an apiary buzzing with bees (although one strain he bred was so nasty-tempered and vindictive it had to be destroyed), and cofounded the Austrian Meteorological Society.

In the early 1860s, just before Miescher moved from medical school into research, Mendel began some deceptively simple experiments on pea plants in the St. Thomas nursery. Beyond enjoying their taste and wanting a ready supply, he chose peas because they simplified experiments. Neither bees nor wind could pollinate his pea blossoms, so he could control which plants mated with which. He appreciated the binary, either/or nature of pea plants, too: plants had tall or short stalks, green or yellow pods, wrinkled or smooth peas, nothing in between. In fact, Mendel's first important conclusion from his work was that some binary traits "dominated" others. For example, crossing purebred green-pead plants with purebred yellow-pead plants produced only yellow-pead offspring: yellow dominated. Importantly, however, the green trait hadn't disappeared. When Mendel mated those second-generation yellow-pead plants with each other, a few furtive green peas popped up—one latent, "recessive" green for every three dominant yellows. The 3:1 ratio* held for other traits, too.

Equally important, Mendel concluded that having one dominant or recessive trait didn't affect whether another, separate trait was dominant or recessive—each trait was independent. For example, even though tall dominated short, a recessive-short plant could still have dominant-yellow peas. Or a tall plant could have recessive-green peas. In fact, every one of the seven traits he studied—like smooth peas (dominant) versus wrinkled peas (recessive), or purple blossoms (dominant) versus white blossoms (recessive)—was inherited independently of the other traits.

This focus on separate, independent traits allowed Mendel to succeed where other heredity-minded horticulturists had failed. Had Mendel tried to describe, all at once, the overall resemblance of a plant to its parents, he would have had too many traits to consider. The plants would have seemed a confusing collage of Mom and Dad. (Charles Darwin, who also grew

---

* This and all upcoming asterisks refer to the Notes and Errata section, which begins on p. 363 and goes into more detail on various interesting points.

and experimented with pea plants, failed to understand their heredity partly for this reason.) But by narrowing his scope to one trait at a time, Mendel could see that each trait must be controlled by a separate factor. Mendel never used the word, but he identified the discrete, inheritable factors we call genes today. Mendel's peas were the Newton's apple of biology.

Beyond his qualitative discoveries, Mendel put genetics on solid quantitative footing. He adored the statistical manipulations of meteorology, the translating of daily barometer and thermometer readings into aggregate climate data. He approached breeding the same way, abstracting from individual plants into general laws of inheritance. In fact, rumors have persisted for almost a century now that Mendel got carried away here, letting his love of perfect data tempt him into fraud.

If you flip a dime a thousand times, you'll get approximately five hundred FDRs and five hundred torches; but you're unlikely to get exactly five hundred of either, because each flip is independent and random. Similarly, because of random deviations, experimental data always stray a tad higher or lower than theory predicts. Mendel should therefore have gotten only approximately a 3:1 ratio of tall to short plants (or whatever other trait he measured). Mendel, however, claimed some almost platonically perfect 3:1s among his thousands of pea plants, a claim that has raised suspicions among modern geneticists. One latter-day fact checker calculated the odds at less than one in ten thousand that Mendel—otherwise a pedant for numerical accuracy in ledgers and meteorological experiments—came by his results honestly. Many historians have defended Mendel over the years or argued that he manipulated his data only unconsciously, since standards for recording data differed back then. (One sympathizer even invented, based on no evidence, an overzealous gardening assistant who knew what numbers Mendel wanted and furtively discarded plants to please his master.) Mendel's original lab notes were burned after his death, so we can't check if he cooked the books. Honestly, though, if Mendel did cheat, it's almost more remarkable: it means he intuited the correct answer—the golden 3:1 ratio of genetics—before having any real proof. The purportedly fraudulent data may simply have been the monk's way of tidying up the vagaries of real-world experiments, to make his data more convincing, so that others could see what he somehow knew by revelation.

Regardless, no one in Mendel's lifetime suspected he'd pulled a fast one—partly because no one was paying attention. He read a paper on pea heredity at a conference in 1865, and as one historian noted, "his audience dealt with him in the way that all audiences do when presented with more mathematics than they have a taste for: there was no discussion, and no questions were asked." He almost shouldn't have bothered, but Mendel published his results in 1866. Again, silence.

Mendel kept working for a few years, but his chance to burnish his scientific reputation largely evaporated in 1868, when his monastery elected him abbot. Never having governed anything before, Mendel had a lot to learn, and the day-to-day headaches of running St. Thomas cut into his free time for horticulture. Moreover, the perks of being in charge, like rich foods and cigars (Mendel smoked up to twenty cigars per day and grew so stout that his resting pulse sometimes topped 120), slowed him down, limiting his enjoyment of the gardens and greenhouses. One later visitor did remember Abbot Mendel taking him on a stroll through the gardens and pointing out with delight the blossoms and ripe pears; but at the first mention of his own experiments in the garden, Mendel changed the subject, almost embarrassed. (Asked how he managed to grow nothing but tall pea plants, Mendel demurred: "It is just a little trick, but there is a long story connected with it, which would take too long to tell.")

Mendel's scientific career also atrophied because he wasted

an increasing number of hours squabbling about political issues, especially separation of church and state. (Although it's not obvious from his scientific work, Mendel could be fiery—a contrast to the chill of Miescher.) Almost alone among his fellow Catholic abbots, Mendel supported liberal politics, but the liberals ruling Austria in 1874 double-crossed him and revoked the tax-exempt status of monasteries. The government demanded seventy-three hundred gulden per year from St. Thomas in payment, 10 percent of the monastery's assessed value, and although Mendel, outraged and betrayed, paid some of the sum, he refused to pony up the rest. In response, the government seized property from St. Thomas's farms. It even dispatched a sheriff to seize assets from inside St. Thomas itself. Mendel met his adversary in full clerical habit outside the front gate, where he stared him down and dared him to extract the key from his pocket. The sheriff left empty-handed.

Overall, though, Mendel made little headway getting the new law repealed. He even turned into something of a crank, demanding interest for lost income and writing long letters to legislators on arcane points of ecclesiastical taxation. One lawyer sighed that Mendel was "full of suspicion, [seeing] himself surrounded by nothing but enemies, traitors, and intriguers." The "Mendel affair" did make the erstwhile scientist famous, or notorious, in Vienna. It also convinced his successor at St. Thomas that Mendel's papers should be burned when he died, to end the dispute and save face for the monastery. The notes describing the pea experiments would become collateral casualties.

Mendel died in 1884, not long after the church-state imbroglio; his nurse found him stiff and upright on his sofa, his heart and kidneys having failed. We know this because Mendel feared being buried alive and had demanded a precautionary autopsy. But in one sense, Mendel's fretting over a premature burial proved prophetic. Just eleven scientists cited his now-classic paper on inheritance in the thirty-five years after his death. And those that did (mostly agricultural scientists) saw his experiments as mildly interesting lessons for breeding peas, not universal statements on heredity. Scientists had indeed buried Mendel's theories too soon.

But all the while, biologists were discovering things about cells that, if they'd only known, supported Mendel's ideas. Most important, they found distinct ratios of traits among offspring, and determined that chromosomes passed hereditary information around in discrete chunks, like the discrete traits Mendel identified. So when three biologists hunting through footnotes around 1900 all came across the pea paper independently and realized how closely it mirrored their own work, they grew determined to resurrect the monk.

Mendel allegedly once vowed to a colleague, "My time will come," and boy, did it. After 1900 "Mendelism" expanded so quickly, with so much ideological fervor pumping it up, that it began to rival Charles Darwin's natural selection as the preeminent theory in biology. Many geneticists in fact saw Darwinism and Mendelism as flatly incompatible—and a few even relished the prospect of banishing Darwin to the same historical obscurity that Friedrich Miescher knew so well.

2

The Near Death of Darwin

Why Did Geneticists Try to Kill Natural Selection?

 his was not how a Nobel laureate should have to spend his time. In late 1933, shortly after winning science's highest honor, Thomas Hunt Morgan got a message from his longtime assistant Calvin Bridges, whose libido had landed him in hot water. Again.

A "confidence woman" from Harlem had met Bridges on a cross-country train a few weeks before. She quickly convinced him not only that she was a regal princess from India, but that her fabulously wealthy maharaja of a father just happened to have opened — coincidence of all coincidences — a science institute on the subcontinent in the very field that Bridges (and Morgan) worked in, fruit fly genetics. Since her father needed a man to head the institute, she offered Bridges the job. Bridges, a real Casanova, would likely have shacked up with the woman anyway, and the job prospect made her irresistible. He was so smitten he began offering his colleagues jobs in India and didn't seem to notice Her Highness's habit of running up extraordinary bills whenever they went carousing. In fact, when out of earshot, the supposed princess claimed to be Mrs. Bridges and charged everything she could to him. When the truth emerged, she tried to extort more cash by threatening to sue him "for transporting her across state lines for immoral purposes." Panicked and distraught—despite his adult activities, Bridges was quite childlike—he turned to Morgan.

Morgan no doubt consulted with his other trusted assistant, Alfred Sturtevant. Like Bridges, Sturtevant had worked with Morgan for decades, and the trio had shared in some of the most important discoveries in genetics history. Sturtevant and Morgan both scowled in private over Bridges's dalliances and escapades, but their loyalty trumped any other consideration here. They decided that Morgan should throw his weight around. In short order, he threatened to expose the woman to the police, and kept up the pressure until Miss Princess disappeared on the next train. Morgan then hid Bridges away until the situation blew over.*

When he'd hired Bridges as a factotum years before, Morgan could never have expected he'd someday be acting as a goodfella for him. Then again, Morgan could never have expected how most everything in his life had turned out. After laboring away in anonymity, he had now become a grand panjandrum of genetics. After working in comically cramped quarters in Manhattan, he now oversaw a spacious lab in California. After lavishing so much attention and affection on his "fly boys" over the years, he was now fending off charges from former assistants that he'd stolen credit for others' ideas. And after fighting so hard for so long against the overreach of ambitious scientific theories, he'd now surrendered to, and even helped expand, the two most ambitious theories in all biology.

Morgan's younger self might well have despised his older self for this last thing. Morgan had begun his career at a curious time in science history, around 1900, when a most uncivil civil war broke out between Mendel's genetics and Darwin's natural selection: things got so nasty, most biologists felt that one theory or the other would have to be exterminated. In this war Morgan had tried to stay Switzerland, refusing at first to accept either theory. Both relied too much on speculation, he felt, and Morgan had an almost reactionary distrust of speculation. If he couldn't see proof for a theory in front of his corneas, he wanted to banish it from science. Indeed, if scientific advances often require a brilliant theorist to emerge and explain his vision with perfect clarity, the opposite was true for Morgan, who was cussedly stubborn and notoriously muddled in his reasoning— anything but literally visible proof bemused him.

And yet that very confusion makes him the perfect guide to follow along behind during the War of the Roses interlude when Darwinists and Mendelists despised each other. Morgan mistrusted genetics and natural selection equally at first, but his patient experiments on fruit flies teased out the half-truths of each. He eventually succeeded—or rather, he and his talented team of assistants succeeded—in weaving genetics and evolution together into the grand tapestry of modern biology.

The decline of Darwinism, now known as the "eclipse" of Darwinism, began in the late 1800s and began for quite rational reasons. Above all, while biologists gave Darwin credit for proving that evolution happened, they disparaged his mechanism for evolution—natural selection, the survival of the fittest—as woefully inadequate for bringing about the changes he claimed.

Critics harped especially on their belief that natural selection merely executed the unfit; it seemed to illuminate nothing about where new or advantageous traits come from. As one wit said, natural selection accounted for the survival, but not the *arrival*, of the fittest. Darwin had compounded the problem by insisting that natural selection worked excruciatingly slowly, on tiny differences among individuals. No one else believed that such minute variations could have any practical long-term difference—they believed in evolution by jerks and jumps. Even Darwin's bulldog Thomas Henry Huxley recalled trying, "much to Mr. Darwin's disgust," to convince Darwin that species sometimes advanced by jumps. Darwin wouldn't budge—he accepted only infinitesimal steps.

Additional arguments against natural selection gathered strength after Darwin died in 1882. As statisticians had demonstrated, most traits for species formed a bell curve: ⌒. Most people stood an average height, for example, and the number of tall or short people dropped smoothly to small numbers on both sides. Traits in animals like speed (or strength or smarts) also formed bell curves, with a large number of average creatures. Obviously natural selection would weed out the slowpokes and idiots when predators snatched them. For evolution to occur, though, most scientists argued that the average had to shift; your average creature had to become faster or stronger or smarter. Otherwise the species largely remained the same. But killing off the slowest creatures wouldn't suddenly make those that escaped any faster—and the escapees would continue having mediocre children as a result. What's more, most scientists assumed that the speed of any rare fast creature would be diluted when it bred with slower ones, producing more mediocrities. According to this logic, species got stuck in ruts of average traits, and the nudge of natural selection couldn't improve them. True evolution, then—men from monkeys—had to proceed by jumps.*

Beyond its apparent statistical problems, Darwinism had something else working against it: emotion. People loathed natural selection. Pitiless death seemed paramount, with superior types always crushing the weak. Intellectuals like playwright George Bernard Shaw even felt betrayed by Darwin. Shaw had adored Darwin at first for smiting religious dogmas. But the more Shaw heard, the less he liked natural selection. And "when its whole significance dawns on you," Shaw later lamented, "your heart sinks into a heap of sand within you. There is a hideous fatalism about it, a ghastly and damnable reduction of beauty and intelligence." Nature governed by such rules, he said, would be "a universal struggle for hogwash."

The triplicate rediscovery of Mendel in 1900 further galvanized the anti-Darwinists by providing a scientific alternative — and soon an outright rival. Mendel's work emphasized not murder and starvation but growth and generation. Moreover, Mendel's peas showed signs of jerkiness — tall or short stalks, yellow or green peas, nothing in between. Already by 1902 the English biologist William Bateson had helped a doctor identify the first known gene in humans (for an alarming but largely benign disorder, alkaptonuria, which can turn children's urine black). Bateson soon rebranded Mendelism "genetics" and became Mendel's bulldog in Europe, tirelessly championing the monk's work, even taking up chess and cigars simply because Mendel loved both. Others supported Bateson's creepy zealotry, however, because Darwinism violated the progressive ethos of the young century. Already by 1904, German scientist Eberhard Dennert could cackle, "We are standing at the death-bed of Darwinism, and making ready to send the friends of the patient a little money, to ensure a decent burial." (A sentiment fit for a creationist today.) To be sure, a minority of biologists defended Darwin's vision of gradual evolution against the Dennerts and Batesons of the world, and defended it fiercely — one historian commented on both sides' "remarkable degree of bitchiness." But these stubborn few could not prevent the eclipse of Darwinism from growing ever darker.

Still, while Mendel's work galvanized the anti-Darwinists, it never quite united them. By the early 1900s, scientists had discovered various important facts about genes and chromosomes, facts that still undergird genetics today. They determined that all creatures have genes; that genes can change, or mutate; that all chromosomes in cells come in pairs; and that all creatures inherit equal numbers of chromosomes from Mom and Dad. But there was no overarching sense of how these discoveries meshed; the individual pixels never resolved into a coherent picture. Instead a baffling array of half theories emerged, like "chromosome theory," "mutation theory," "gene theory," and so on. Each championed one narrow aspect of heredity, and each drew distinctions that seem merely confusing today: some scientists believed (wrongly) that genes didn't reside on chromosomes; others that each chromosome harbored just one gene; still others that chromosomes played no role in heredity at all. It's whiggishly unfair to say, but reading these overlapping theories can be downright frustrating today. You want to scream at the scientists, like a dimwit on *Wheel of Fortune* or something, "Think! It's all right there!" But each fiefdom discounted discoveries by rivals, and they squabbled against each other almost as much as against Darwinism.

As these revolutionaries and counterrevolutionaries bitched it out in Europe, the scientist who eventually ended the Darwin-genetics row was working in anonymity in America. Though he mistrusted both Darwinists and geneticists—too much bloviating about theory all around—Thomas Hunt Morgan had developed an interest in heredity after visiting a botanist in Holland in 1900. Hugo de Vries had been one of the trio who rediscovered Mendel that year, and de Vries's fame in Europe rivaled Darwin's, partly because de Vries had developed a rival theory for the origin of species. De Vriesian "mutation theory" argued that species went through rare but intense mutation periods, during which the parents produced "sports," offspring with markedly different traits. De Vries developed mutation theory after spotting some anomalous evening primroses in an abandoned potato field near Amsterdam. Some of these sport primroses sported smoother leaves, longer stems, or bigger yellow flowers with more petals. And crucially, primrose sports wouldn't mate with the old, normal primroses; they seemed to have jumped past them and become a new species. Darwin had rejected evolutionary jumps because he believed that if one sport emerged, it would have to breed with normal individuals, diluting its good qualities. De Vries's mutation period removed this objection at a stroke: many sports emerged at once, and they could breed only with each other.

The primrose results scored themselves into Morgan's brain. That de Vries had no clue how or why mutations appeared mattered not a lick. At last Morgan saw proof of new species emerging, not speculation. After landing a post at Columbia University in New York, Morgan decided to study mutation periods in animals. He began experiments on mice, guinea pigs, and pigeons, but when he discovered how slowly they bred, he took a colleague's suggestion and tried *Drosophila*, fruit flies.

Like many New Yorkers then, fruit flies had recently immigrated, in their case arriving on boats with the first banana crops in the 1870s. These exotic yellow fruits, usually wrapped in foil, had sold for ten cents per, and guards in New York stood watch over banana trees to prevent eager mobs from stealing the fruit. But by 1907 bananas and flies were common enough in New York that Morgan's assistant could catch a whole horde for research simply by slicing up a banana and leaving it on a windowsill to rot.

Fruit flies proved perfect for Morgan's work. They bred quickly—one generation every twelve days—and survived on food cheaper than peanuts. They also tolerated claustrophobic Manhattan real estate. Morgan's lab—the "fly room," 613 Schermerhorn Hall at Columbia—measured sixteen feet by

Thomas Hunt Morgan's cluttered, squalid fly room at Columbia University. Hundreds of fruit flies swarmed around inside each bottle, surviving on rotten bananas. (The American Philosophical Society)

twenty-three feet and had to accommodate eight desks. But a thousand fruit flies lived happily in a one-quart milk bottle, and Morgan's shelves were soon lined with the dozens of bottles that (legend has it) his assistants "borrowed" from the student cafeteria and local stoops.

Morgan set himself up at the fly room's central desk. Cockroaches scuttled through his drawers, nibbling rotten fruit, and the room was a cacophony of buzzing, but Morgan stood unperturbed in the middle of all, peering through a jeweler's loupe, scrutinizing bottle after bottle for de Vries's mutants. When a bottle produced no interesting specimens, Morgan might squash them with his thumb and smear their guts wherever, often in lab notebooks. Unfortunately for general sanitation, Morgan had many, many flies to smush: although the *Drosophila* bred and bred and bred, he found no sign of sports.

Meanwhile Morgan got lucky in a different arena. In autumn 1909, he filled in for a colleague on sabbatical and taught the only introductory course of his Columbia career. And during that semester he made, one observer noted, "his greatest discovery," two brilliant assistants. Alfred Sturtevant heard about Morgan's class through a brother who taught Latin and Greek at Columbia, and although just a sophomore, Sturtevant impressed Morgan by writing an independent research paper on horses and the inheritance of coat color. (Morgan hailed from Kentucky, and his father and uncle had been famous horse thieves behind Union lines during the Civil War, leading a band known as Morgan's Raiders. Morgan scorned his Confederate past, but he knew his horses.) From that moment on, Sturtevant was Morgan's golden boy and eventually earned a coveted desk in the fly room. Sturtevant cultivated a cultured air, reading widely in literature and doing trickier British crossword puzzles—although, amid the fly room's squalor, someone also once discovered a mummified mouse in his desk. Sturtevant did have one deficiency as a scientist, red-green color blindness. He'd tended horses on the family fruit farm in Alabama largely because he proved useless during harvest, struggling to spot red strawberries on green bushes.

The other undergrad, Calvin Bridges, made up for Sturtevant's poor eyesight, and his stuffiness. At first Morgan merely took pity on Bridges, an orphan, by giving him a job washing filth from milk bottles. But Bridges eavesdropped on Morgan's discussions of his work, and when Bridges began spotting interesting flies with his bare eyes (even through the dirty glass bottles), Morgan hired him as a researcher. It was basically the only job Bridges ever had. A sensuous and handsome man with bouffant hair, Bridges practiced free love *avant la lettre*. He eventually abandoned his wife and children, got a vasectomy, and started brewing moonshine in his new bachelor lair in Manhattan.

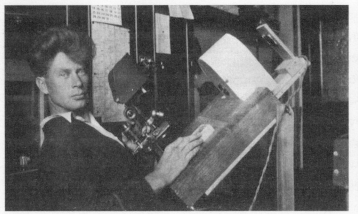

Playboy Calvin Bridges (*left*) and a rare photo of Thomas Hunt Morgan (*right*). Morgan so detested having his picture taken that an assistant who once wanted one had to hide a camera in a bureau in the fly lab and snap the photo remotely by tugging a string. (Courtesy of the National Library of Medicine)

He proceeded to hit on—or flat-out proposition—anything in a skirt, including colleagues' wives. His naive charm seduced many, but even after the fly room became legendary, no other university would blacken its reputation by employing Bridges as anything but a measly assistant.

Meeting Bridges and Sturtevant must have cheered Morgan, because his experiments had all but flopped until then. Unable to find any natural mutants, he'd exposed flies to excess heat and cold and injected acids, salts, bases, and other potential mutagens into their genitals (not easy to find). Still nothing. On the verge of giving up, in January 1910 he finally spotted a fly with a strange trident shape tattooed on its thorax. Not exactly a de Vriesian über-fly, but something. In March two more mutants appeared, one with ragged moles near its wings that made it appear to have "hairy armpits," another with an olive (instead of the normal amber) body color. In May 1910 the most dramatic mutant yet appeared, a fly with white (instead of red) eyes.

Anxious for a breakthrough—perhaps this was a mutation period—Morgan tediously isolated white-eyes. He uncapped the milk bottle, balanced another one upside down on top of it like mating ketchup bottles, and shined a light through the top to coax white-eyes upward. Of course, hundreds of other flies joined white-eyes in the top bottle, so Morgan had to quickly cap both, get a new milk bottle, and repeat the process over and over, slowly dwindling the number with each step, praying to God white-eyes didn't escape meantime. When he finally, finally segregated the bug, he mated it with red-eyed females. Then he bred the descendants with each other in various ways. The results were complex, but one result especially excited Morgan: after crossing some red-eyed descendants with each other, he discovered among the offspring a 3:1 ratio of red to white eyes.

The year before, in 1909, Morgan had heard the Danish botanist Wilhelm Johannsen lecture about Mendelian ratios at Columbia. Johannsen used the occasion to promote his newly minted word, *gene*, a proposed unit of inheritance. Johannsen and others freely admitted that genes were convenient fictions, linguistic placeholders for, well, something. But they insisted that their ignorance about the biochemical details of genes shouldn't invalidate the usefulness of the gene concept for study-ing inheritance (similar to how psychologists today can study euphoria or depression without understanding the brain in detail). Morgan found the lecture too speculative, but his experimental results—3:1—promptly lowered his prejudice to Mendel.

This was quite a volte-face for Morgan, but it was just the start. The eye-color ratios convinced him that gene theory wasn't bunk. But where were genes actually located? Perhaps on chromosomes, but fruit flies had hundreds of inheritable traits and only four chromosomes. Assuming one trait per chromo-some, as many scientists did, there weren't enough to go around. Morgan didn't want to get dragged into debates on so-called chromosome theory, but a subsequent discovery left him no choice: because when he scrutinized his white-eyed flies, he discovered that every last mutant was male. Scientists already knew that one chromosome determined the gender of flies. (As in mammals, female flies have two X chromosomes, males one.) Now the white-eye gene was linked to that chromosome as well—putting two traits on it. Soon the fly boys found other genes—stubby wings, yellow bodies—also linked exclusively to males. The conclusion was inescapable: they'd proved that multiple genes rode around together on one chromosome.* That Morgan had proved this practically against his own will mattered little; he began to champion chromosome theory anyway.

Overthrowing old beliefs like this became a habit with Morgan, simultaneously his most admirable and most maddening trait. Although he encouraged theoretical discussions in the fly room, Morgan considered new theories cheap and facile—worth little until cross-examined in the lab. He didn't seem to grasp that scientists need theories as guides, to decide what's relevant and what's not, to frame their results and prevent muddled thinking. Even undergraduates like Bridges and Sturtevant—and especially a student who joined the fly room later, the abrasively brilliant and brilliantly abrasive Hermann Muller—grew hair-rippingly frustrated with Morgan in the many quarrels they had over genes and heredity. And then, just as exasperating, when someone did wrestle Morgan into a headlock and convince him he was wrong, Morgan would ditch his old ideas and with no embarrassment whatsoever absorb the new ones as obvious.

To Morgan, this quasi plagiarism was no big deal. Everyone was working toward the same goal (*right*, fellas?), and only experiments mattered anyway. And to his credit, his about-faces proved that Morgan listened to his assistants, a contrast to the condescending relationship most European scientists had with their help. For this reason Bridges and Sturtevant always publicly professed their loyalty to Morgan. But visitors sometimes picked up on sibling rivalries among the assistants, and secret smoldering. Morgan didn't mean to connive or manipulate; credit for ideas just meant that little to him.

Nevertheless ideas kept ambushing Morgan, ideas he hated. Because not long after the unified gene-chromosome theory emerged, it nearly unraveled, and only a radical idea could salvage it. Again, Morgan had determined that multiple genes clustered together on one chromosome. And he knew from other scientists' work that parents pass whole chromosomes on to their children. All the genetic traits on each chromosome should therefore always be inherited together—they should always be linked. To take a hypothetical example, if one chromosome's set of genes call for green bristles and sawtooth wings and fat antennae, any fly with one trait should exhibit all three. Such clusters of traits do exist in flies, but to their dismay, Morgan's team discovered that certain linked traits could sometimes become unlinked—green bristles and sawtooth wings, which should always appear together, would somehow show up separately, in different flies. Unlinkings weren't common—linked traits might separate 2 percent of the time, or 4 percent—but they were so persistent they might have undone the entire theory, if Morgan hadn't indulged in a rare flight of fancy.

He remembered reading a paper by a Belgian biologist-priest who had used a microscope to study how sperm and eggs form. One key fact of biology—it comes up over and over—is that all chromosomes come in pairs, pairs of nearly identical twins. (Humans have forty-six chromosomes, arranged in twenty-three pairs.) When sperm and eggs form, these near-twin chromosomes all line up in the middle of the parent cell. During division one twin gets pulled one way, the other the other way, and two separate cells are born.

However, the priest-biologist noticed that, just before the

divvying up, twin chromosomes sometimes interacted, coiling their tips around each other. He didn't know why. Morgan suggested that perhaps the tips broke off during this crossing over and swapped places. This explained why linked traits sometimes separated: the chromosome had broken somewhere between the two genes, dislocating them. What's more, Morgan speculated — he was on a roll — that traits separating 4 percent of the time probably sat farther apart on chromosomes than those separating 2 percent of the time, since the extra distance between the first pair would make breaking along that stretch more likely.

Morgan's shrewd guess turned out correct, and with Sturtevant and Bridges adding their own insights over the next few years, the fly boys began to sketch out a new model of heredity — the model that made Morgan's team so historically important. It said that all traits were controlled by genes, and that these genes resided on chromosomes in fixed spots, strung along like pearls on a necklace. Because creatures inherit one copy of each chromosome from each parent, chromosomes therefore pass genetic traits from parent to child. Crossing over (and mutation) changes chromosomes a little, which helps make each creature unique. Nevertheless chromosomes (and genes) stay mostly intact, which explains why traits run in families. Voilà: the first overarching sense of how heredity works.

In truth, little of this theory originated in Morgan's lab, as biologists worldwide had discovered various pieces. But Morgan's team finally linked these vaguely connected ideas, and fruit flies provided overwhelming experimental proof. No one could deny that sex chromosome linkage occurred, for instance, when Morgan had ten thousand mutants buzzing on a shelf, nary a female among them.

Of course, while Morgan won acclaim for uniting these theories, he'd done nothing to reconcile them with Darwinian natural selection. That reconciliation also arose from work inside the fly room, but once again Morgan ended up "borrowing" the idea from assistants, including one who didn't accept this as docilely as Bridges and Sturtevant did.

Hermann Muller began poking around the fly room in 1910, though only occasionally. Because he supported his elderly mother, Muller lived a haphazard life, working as a factotum in hotels and banks, tutoring immigrants in English at night, bolting down sandwiches on the subway between jobs. Somehow Muller found time to befriend writer Theodore Dreiser in Greenwich Village, immerse himself in socialist politics, and commute two hundred miles to Cornell University to finish a master's degree. But no matter how frazzled he got, Muller used his one free day, Thursday, to drop in on Morgan and the fly boys and bandy about ideas on genetics. Intellectually nimble, Muller starred in these bull sessions, and Morgan granted him a desk in the fly room after he graduated from Cornell in 1912. The problem was, Morgan declined to pay Muller, so Muller's schedule didn't let up. He soon had a mental breakdown.

From then on, and for decades afterward, Muller seethed over his status in the fly room. He seethed that Morgan openly favored the bourgeois Sturtevant and shunted menial tasks like preparing bananas onto the blue-collar, proletariat Bridges. He seethed that both Bridges and Sturtevant got paid to experiment on his, Muller's, ideas, while he scrambled around the five boroughs for pocket change. He seethed that Morgan treated the fly room like a clubhouse and sometimes made Muller's friends work down the hall. Muller seethed above all that Morgan was oblivious to his contributions. This was partly because Muller proved slow in doing the thing Morgan most valued—actually carrying out the clever experiments he (Muller) dreamed up. Indeed, Muller probably couldn't have found a worse mentor than Morgan. For all his socialist leanings, Muller got pretty attached to his own intellectual property, and felt the free and communal nature of the fly room both exploited and ignored his talent. Nor was Muller exactly up for Mr. Congeniality. He harped on Morgan, Bridges, and Sturtevant with tactless criticism, and got almost personally offended by anything but pristine logic. Morgan's breezy dismissal of evolution by natural selection especially irked Muller, who considered it the foundation of biology.

Despite the personality clashes he caused, Muller pushed the fly group to greater work. In fact, while Morgan contributed little to the emerging theory of inheritance after 1911, Muller, Bridges, and Sturtevant kept making fundamental discoveries. Unfortunately, it's hard to sort out nowadays who discovered what, and not just because of the constant idea swapping. Morgan and Muller often scribbled thoughts down on unorganized scraps, and Morgan purged his file cabinet every five years, perhaps out of necessity in his cramped lab. Muller hoarded documents, but many years later, yet another colleague he'd managed to alienate threw out Muller's files while Muller was working abroad. Morgan also (like Mendel's fellow friars) destroyed Bridges's files when the free lover died of heart problems in 1938. Turns out Bridges was a bedpost notcher, and when Morgan found a detailed catalog of fornication, he thought it prudent to burn all the papers and protect everyone in genetics.

But historians can assign credit for some things. All the fly boys helped determine which clusters of traits got inherited together. More important, they discovered that four distinct clusters existed in flies—exactly the number of chromosome pairs. This was a huge boost for chromosome theory because it showed that every chromosome harbored multiple genes.

Sturtevant built on this notion of gene and chromosome linkage. Morgan had guessed that genes separating 2 percent of the time must sit closer together on chromosomes than genes separating 4 percent of the time. Ruminating one evening, Sturtevant realized he could translate those percentages into actual distances. Specifically, genes separating 2 percent of the time must sit twice as close together as the other pair; similar logic held for other percent linkages. Sturtevant blew off his undergraduate homework that night, and by dawn this nineteen-year-old had sketched the first map of a chromosome. When Muller saw the map, he "literally jumped with excitement"—then pointed out ways to improve it.

Bridges discovered "nondisjunction"—the occasional failure of chromosomes to separate cleanly after crossing over and twisting arms. (The excess of genetic material that results can cause problems like Down syndrome.) And beyond individual discoveries, Bridges, a born tinkerer, industrialized the fly room. Instead of tediously separating flies by turning bottle after bottle upside down, Bridges invented an atomizer to puff wee doses of ether over flies and stun them. He also replaced loupes with binocular microscopes; handed out white porcelain plates and fine-tipped paintbrushes so that people could see and manipulate flies more easily; eliminated rotting bananas for a nutritious slurry of molasses and cornmeal; and built climate-controlled cabinets so that flies, which become sluggish in cold, could breed summer and winter. He even built a fly morgue to dispose of corpses with dignity. Morgan didn't always appreciate these contributions—he continued to squish flies wherever they landed, despite the morgue. But Bridges knew that mutants popped up so rarely, and when they did, his biological factory allowed each one to thrive and produce millions of descendants.*

Muller contributed insights and ideas, dissolving apparent contradictions and undergirding lean-to theories with firm logic. And although he had to argue with Morgan until his tongue bled, he finally made the senior scientist see how genes, mutations, and natural selection work together. As Muller (among others) outlined it: Genes give creatures traits, so mutations to genes change traits, making creatures different in color, height, speed, or whatever. But contra de Vries—who saw mutations as large things, producing sports and instant species—most mutations simply tweak creatures. Natural selection then allows the better-adapted of these creatures to survive and reproduce more often. Crossing over comes into play because it shuffles genes around between chromosomes and therefore puts new versions of genes together, giving natural selection still more variety to work on. (Crossing over is so important that some scientists today think that sperm and eggs refuse to form unless chromosomes cross a minimum number of times.)

Muller also helped expand scientists' very ideas about what genes could do. Most significantly, he argued that traits like the ones Mendel had studied—binary traits, controlled by one gene—weren't the only story. Many important traits are controlled by multiple genes, even dozens of genes. These traits will therefore show gradations, depending on which exact genes a creature inherits. Certain genes can also turn the volume up or down on other genes, crescendos and decrescendos that produce still finer gradations. Crucially, however, because genes are discrete and particulate, a beneficial mutation will *not* be diluted between generations. The gene stays whole and intact, so superior parents can breed with inferior types and still pass the gene along.

To Muller, Darwinism and Mendelism reinforced each other beautifully. And when Muller finally convinced Morgan of this, Morgan became a Darwinian. It's easy to chuckle over this—yet another Morgan conversion—and in later writings, Morgan still emphasizes genetics as more important than natural selection. However, Morgan's endorsement was important in a larger sense. Grandiloquent theories (including Darwin's) dominated biology at the time, and Morgan had helped keep the field grounded, always demanding hard evidence. So other biologists knew that if some theory convinced even Thomas Hunt Morgan, it had something going for it. What's more, even Muller recognized Morgan's personal influence. "We should not forget," Muller once admitted, "the guiding personality of Morgan, who infected all the others by his own example—his indefatigable activity, his deliberation, his jolliness, and courage." In the end, Morgan's bonhomie did what Muller's brilliant sniping couldn't: convinced geneticists to reexamine their prejudice against Darwin, and take the proposed synthesis of Darwin and Mendel, natural selection and genetics, seriously.

Many other scientists did indeed take up the work of Morgan's team in the 1920s, spreading the unassuming fruit fly to labs around the world. It soon became the standard animal in genetics, allowing scientists everywhere to compare discoveries on equal terms. Building on such work, a generation of mathematically minded biologists in the 1930s and 1940s began investigating how mutations spread in natural populations, outside the lab. They demonstrated that if a gene gave some creatures even a small survival advantage, that boost could, if compounded long enough, push species in new directions. What's more, most changes would take place in tiny steps, exactly as Darwin had insisted. If the fly boys' work finally showed how to link Mendel with Darwin, these later biologists made the case as rigorous as a Euclidean proof. Darwin had once moaned how "repugnant" math was to him, how he struggled with most anything beyond taking simple measurements. In truth, mathematics buttressed Darwin's theory and ensured his reputation would never lapse again.* And in this way the so-called eclipse of Darwinism in the early 1900s proved exactly that: a period of darkness and confusion, but a period that ultimately passed.

Beyond the scientific gains, the diffusion of fruit flies around the world inspired another legacy, a direct outgrowth of Morgan's "jolliness." Throughout genetics, the names of most genes are ugly abbreviations, and they stand for monstrous freak words that maybe six people worldwide understand. So when discussing, say, the *alox12b* gene, there's often no point in spelling out its name (arachidonate 12-lipoxygenase, 12R type), since doing so confuses rather than clarifies, methinks. (To save everyone a migraine, from now on I'll just state gene acronyms and pretend they stand for nothing.) In contrast, whereas gene names are intimidatingly complex, chromosome names are stupefyingly banal. Planets are named after gods, chemical elements after myths, heroes, and great cities. Chromosomes were named with all the creativity of shoe sizes. Chromosome one is the longest, chromosome two the second longest, and (yawn) so on. Human chromosome twenty-one is actually shorter than chromosome twenty-two, but by the time scientists figured this out, chromosome twenty-one was famous, since having an extra number twenty-one causes Down syndrome. And really, with such boring names, there was no point in fighting over them and bothering to change.

Fruit fly scientists, God bless 'em, are the big exception. Morgan's team always picked sensibly descriptive names for mutant genes like *speck*, *beaded*, *rudimentary*, *white*, and *abnormal*. And this tradition continues today, as the names of most fruit fly genes eschew jargon and even shade whimsical. Different fruit fly genes include *groucho*, *smurf*, *fear of intimacy*, *lost in space*, *smellblind*, *faint sausage*, *tribble* (the multiplying fuzzballs on *Star Trek*), and *tiggywinkle* (after Mrs. Tiggy-winkle, a character from Beatrix Potter). The *armadillo* gene, when mutated, gives fruit flies a plated exoskeleton. The *turnip* gene makes flies stupid. *Tudor* leaves males (as with Henry VIII) childless. *Cleopatra* can kill flies when it interacts with another gene, *asp*. *Cheap date* leaves flies exceptionally tipsy after a sip of alcohol. Fruit fly sex especially seems to inspire clever names. *Ken and barbie* mutants have no genitalia. Male *coitus interruptus* mutants spend just ten minutes having sex (the norm is twenty), while *stuck* mutants cannot physically disengage after coitus. As for females, *dissatisfaction* mutants never have sex at all—they spend all their energy shooing suitors away by snapping their wings. And thankfully, this whimsy with names has inspired the occasional zinger in other areas of genetics. A gene that gives mammals extra nipples earned the name *scaramanga*, after the James Bond villain with too many. A gene that removes blood cells from circulation in fish became the tasteful *vlad tepes*, after Vlad the Impaler, the historical inspiration for Dracula. The backronym for the "POK erythroid myeloid ontogenic" gene in mice—*pokemon*—nearly provoked a lawsuit, since the *pokemon* gene (now known, sigh, as *zbtb7*) contributes to the spread of cancer, and the lawyers for the Pokémon media empire didn't want their cute little pocket monsters confused with tumors. But my winner for the best, and freakiest, gene name goes to the flour beetle's *medea*, after the ancient Greek mother who committed infanticide. *Medea* encodes a protein with the curious property that it's both a poison and its own antidote. So if a mother has this gene but doesn't pass it to an embryo, her body exterminates the fetus—nothing she can do about it. If the fetus has the gene, s/he creates the antidote and lives. (*Medea* is a "selfish genetic element," a gene that demands its own propagation above all, even to the detriment of a creature as a whole.) If you can get beyond the horror, it's a name worthy of the Columbia fruit fly tradition, and it's fitting that the most important clinical work on *medea*—which could lead to very smart insecticides—came after scientists introduced it into *Drosophila* for further study.

But long before these cute names emerged, and even before fruit flies had colonized genetics labs worldwide, the original fly group at Columbia had disbanded. Morgan moved to the California Institute of Technology in 1928 and took Bridges and Sturtevant with him to his new digs in sunny Pasadena. Five years later Morgan became the first geneticist to win the Nobel Prize, "for establishing," one historian noted, "the very principles of genetics he had set out to refute." The Nobel committee has an arbitrary rule that three people at most can share a Nobel, so the committee awarded it to Morgan alone, rather than—as it should have—splitting it between him, Bridges, Sturtevant, and Muller. Some historians argue that Sturtevant did work important enough to win his own Nobel but that his devotion to Morgan and willingness to relinquish credit for ideas diminished his chances. Perhaps in tacit acknowledgment of this, Morgan shared his prize money from the Nobel with Sturtevant and Bridges, setting up college funds for their children. He shared nothing with Muller.

Muller had fled Columbia for Texas by then. He started in 1915 as a professor at Rice University (whose biology department was chaired by Julian Huxley, grandson of Darwin's bulldog) and eventually landed at the University of Texas. Although Morgan's warm recommendation had gotten him the Rice job, Muller actively promoted a rivalry between his Lone Star and Morgan's Empire State groups, and whenever the Texas group made a significant advance, which they trumpeted as a "home run," they preened. In one breakthrough, biologist Theophilus Painter discovered the first chromosomes—inside fruit fly spit glands*—that were large enough to inspect visually, allowing scientists to study the physical basis of genes. But as important as Painter's work was, Muller hit the grand slam in 1927 when he discovered that pulsing flies with radiation would increase their mutation rate by 150 times. Not only did this have health implications, but scientists no longer had to sit around and wait for mutations to pop up. They could mass-produce them. The discovery gave Muller the scientific standing he deserved—and knew he deserved.

Inevitably, though, Muller got into spats with Painter and

other colleagues, then outright brawls, and he soured on Texas. Texas soured on him, too. Local newspapers outed him as a political subversive, and the precursor to the FBI put him under surveillance. Just for fun, his marriage crumbled, and one evening in 1932 his wife reported him missing. A posse of colleagues later found him muddied and disheveled in the woods, soaked by a night of rain, his head still foggy from the barbiturates he'd swallowed to kill himself.

Burned out, humiliated, Muller abandoned Texas for Europe. There he did a bit of a Forrest Gump tour of totalitarian states. He studied genetics in Germany until Nazi goons vandalized his institute. He fled to the Soviet Union, where he lectured Joseph Stalin himself on eugenics, the quest to breed superior human beings through science. Stalin was not impressed, and Muller scurried to leave. To avoid being branded a "bourgeois reactionary deserter," Muller enlisted on the communist side in the Spanish Civil War, working at a blood bank. His side lost, and fascism descended.

Disillusioned yet again, Muller crawled back to the United States, to Indiana, in 1940. His interest in eugenics grew; he later helped establish what became the Repository for Germinal Choice, a "genius sperm bank" in California. And as the capstone to his career, Muller won his own unshared Nobel Prize in 1946 for the discovery that radiation causes genetic mutations. The award committee no doubt wanted to make up for shutting Muller out in 1933. But he also won because the atomic bomb attacks on Hiroshima and Nagasaki in 1945 — which rained nuclear radiation on Japan — made his work sickeningly relevant. If the fly boys' work at Columbia had proved that genes existed, scientists now had to figure out how genes worked and how, in the deadly light of the bomb, they too often failed.

3

Them's the DNA Breaks

How Does Nature Read—and Misread—DNA?

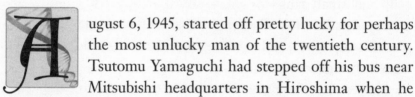ugust 6, 1945, started off pretty lucky for perhaps the most unlucky man of the twentieth century. Tsutomu Yamaguchi had stepped off his bus near Mitsubishi headquarters in Hiroshima when he realized he'd forgotten his *inkan*, the seal that Japanese salary-men dip in red ink and use to stamp documents. The lapse annoyed him—he faced a long ride back to his boardinghouse—but nothing could really dampen his mood that day. He'd finished designing a five-thousand-ton tanker ship for Mitsubishi, and the company would finally, the next day, send him back home to his wife and infant son in southwest Japan. The war had disrupted his life, but on August 7 things would return to normal.

As Yamaguchi removed his shoes at his boardinghouse door, the elderly proprietors ambushed him and asked him to tea. He could hardly refuse these lonely folk, and the unexpected engagement further delayed him. Shod again, *inkan* in hand, he hurried off, caught a streetcar, disembarked near work, and was walking along near a potato field when he heard a gnat of an enemy bomber high above. He could just make out a speck descending from its belly. It was 8:15 a.m.

Many survivors remember the curious delay. Instead of a normal bomb's simultaneous flash-bang, this bomb flashed and swelled silently, and got hotter and hotter silently. Yamaguchi was close enough to the epicenter that he didn't wait long. Drilled in air-raid tactics, he dived to the ground, covered his eyes, and plugged his ears with his thumbs. After a half-second light bath came a roar, and with it came a shock wave. A moment later Yamaguchi felt a gale somehow *beneath* him, raking his stomach. He'd been tossed upward, and after a short flight he hit the ground, unconscious.

He awoke, perhaps seconds later, perhaps an hour, to a darkened city. The mushroom cloud had sucked up tons of dirt and ash, and small rings of fire smoked on wilted potato leaves nearby. His skin felt aflame, too. He'd rolled up his shirtsleeves after his cup of tea, and his forearms felt severely sunburned. He rose and staggered through the potato field, stopping every few feet to rest, shuffling past other burned and bleeding and torn-open victims. Strangely compelled, he reported to Mitsubishi. He found a pile of rubble speckled with small fires, and many dead coworkers—he'd been lucky to be late. He wandered onward; hours slipped by. He drank water from broken pipes, and at an emergency aid station, he nibbled a biscuit and vomited. He slept that night beneath an overturned boat on a beach. His left arm, fully exposed to the great white flash, had turned black.

All the while, beneath his incinerated skin, Yamaguchi's DNA was nursing even graver injuries. The nuclear bomb at Hiroshima released (among other radioactivity) loads of super-charged x-rays called gamma rays. Like most radioactivity, these rays single out and selectively damage DNA, punching DNA and nearby water molecules and making electrons fly out like uppercut teeth. The sudden loss of electrons forms free radicals, highly reactive atoms that chew on chemical bonds. A chain reaction begins that cleaves DNA and sometimes snaps chromosomes into pieces.

By the mid-1940s, scientists were starting to grasp why the shattering or disruption of DNA could wreak such ruin inside cells. First, scientists based in New York produced strong evidence that genes were made of DNA. This upended the persistent belief in protein inheritance. But as a second study revealed, DNA and proteins still shared a special relationship: DNA *made* proteins, with each DNA gene storing the recipe for one protein. Making proteins, in other words, was what genes did—that's how genes created traits in the body.

In conjunction, these two ideas explained the harm of radioactivity. Fracturing DNA disrupts genes; disrupting genes halts protein production; halting protein production kills cells. Scientists didn't work this out instantly—the crucial "one gene/one protein" paper appeared just days before Hiroshima—but they knew enough to cringe at the thought of nuclear weapons. When Hermann Muller won his Nobel Prize in 1946, he prophesied to the *New York Times* that if atomic bomb survivors "could foresee the results 1,000 years from now...they might consider themselves more fortunate if the bomb had killed them."

Despite Muller's pessimism, Yamaguchi did want to survive, badly, for his family. He'd had complicated feelings about the war—opposing it at first, supporting it once under way, then shading back toward opposition when Japan began to stumble, because he feared the island being overrun by enemies who might harm his wife and son. (If so, he'd contemplated giving them an overdose of sleeping pills to spare them.) In the hours after Hiroshima, he yearned to get back to them, so when he heard rumors about trains leaving the city, he sucked up his strength and resolved to find one.

Hiroshima is a collection of islands, and Yamaguchi had to

cross a river to reach the train station. All the bridges had collapsed or burned, so he steeled himself and began crossing an apocalyptic "bridge of corpses" clogging the river, crawling across melted legs and faces. But an uncrossable gap in the bridge forced him to turn back. Farther upstream, he found a railroad trestle with one steel beam intact, spanning fifty yards. He clambered up, crossed the iron tightrope, and descended. He pushed through the mob at the station and slumped into a train seat. Miraculously the train pulled out soon afterward—he was saved. The train would run all night, but he was finally headed home, to Nagasaki.

A physicist stationed in Hiroshima might have pointed out that the gamma rays finished working over Yamaguchi's DNA in a millionth of a billionth of a second. To a chemist, the most interesting part—how the free radicals gnawed through DNA—would have ceased after a millisecond. A cell biologist would have needed to wait maybe a few hours to study how cells patch up torn DNA. A doctor could have diagnosed radiation sickness—headaches, vomiting, internal bleeding, peeling skin, anemic blood—within a week. Geneticists needed the most patience. The genetic damage to the survivors didn't surface for years, even decades. And in an eerie coincidence, scientists began to piece together how exactly genes function, and fail, during those very decades—as if providing a protracted running commentary on DNA devastation.

However definitive in retrospect, experiments on DNA and proteins in the 1940s convinced only some scientists that DNA was the genetic medium. Better proof came in 1952, from virologists Alfred Hershey and Martha Chase. Viruses, they knew, hijacked cells by injecting genetic material. And because the viruses they studied consisted of only DNA and proteins, genes

had to be one or the other. The duo determined which by tag-
ging viruses with both radioactive sulfur and radioactive phosphorus, then turning them loose on cells. Proteins contain
sulfur but no phosphorus, so if genes were proteins, radioactive
sulfur should be present in cells postinfection. But when Hershey and Chase filtered out infected cells, only radioactive phosphorus remained: only DNA had been injected.

Hershey and Chase published these results in April 1952,
and they ended their paper by urging caution: "Further chemical inferences should not be drawn from the experiments presented." Yeah, right. Every scientist in the world still working on
protein heredity dumped his research down the sink and took up
DNA. A furious race began to understand the structure of DNA,
and just one year later, in April 1953, two gawky scientists at
Cambridge University in England, Francis Crick and James
Watson (a former student of Hermann Muller), made the term
"double helix" legendary.

Watson and Crick's double helix was two loooooooong DNA
strands wrapped around each other in a right-handed spiral.
(Point your right thumb toward the ceiling; DNA twists upward
along the counterclockwise curl of your fingers.) Each strand
consisted of two backbones, and the backbones were held together
by paired bases that fit together like puzzle pieces—angular A
with T, curvaceous C with G. Watson and Crick's big insight
was that because of this complementary A-T and C-G base pairing, one strand of DNA can serve as a template for copying the
other. So if one side reads CCGAGT, the other side must read
GGCTCA. It's such an easy system that cells can copy hundreds
of DNA bases per second.

However well hyped, though, the double helix revealed zero
about how DNA genes actually made proteins—which is, after
all, the important part. To understand this process, scientists
had to scrutinize DNA's chemical cousin, RNA. Though similar

to DNA, RNA is single-stranded, and it substitutes the letter U (uracil) for T in its strands. Biochemists focused on RNA because its concentration would spike tantalizingly whenever cells started making proteins. But when they chased the RNA around the cell, it proved as elusive as an endangered bird; they caught only glimpses before it vanished. It took years of patient experiments to determine exactly what was going on here—exactly how cells transform strings of DNA letters into RNA instructions and RNA instructions into proteins.

Cells first "transcribe" DNA into RNA. This process resembles the copying of DNA, in that one strand of DNA serves as a template. So the DNA string CCGAGT would become the RNA string GGCUCA (with U replacing T). Once constructed, this RNA string leaves the confines of the nucleus and chugs out to special protein-building apparatuses called ribosomes. Because it carries the message from one site to another, it's called messenger RNA.

The protein building, or translation, begins at the ribosomes. Once the messenger RNA arrives, the ribosome grabs it near the end and exposes just three letters of the string, a triplet. In our example, GGC would be exposed. At this point a second type of RNA, called transfer RNA, approaches. Each transfer RNA has two key parts: an amino acid trailing behind it (its cargo to transfer), and an RNA triplet sticking off its prow like a masthead. Various transfer RNAs might try to dock with the messenger RNA's exposed triplet, but only one with complementary bases will stick. So with the triplet GGC, only a transfer RNA with CCG will stick. And when it does stick, the ribosome unloads its amino acid cargo.

At this point the transfer RNA leaves, the messenger RNA shifts down three spots, and the process repeats. A different triplet is exposed, and a different transfer RNA with a different amino acid docks. This puts amino acid number two in place.

Eventually, after many iterations, this process creates a string of amino acids—a protein. And because each RNA triplet leads to one and only one amino acid being added, information should (should) get translated perfectly from DNA to RNA to protein. This same process runs every living thing on earth. Inject the same DNA into guinea pigs, frogs, tulips, slime molds, yeast, U.S. congressmen, whatever, and you get identical amino acid chains. No wonder that in 1958 Francis Crick elevated the DNA → RNA → protein process into the "Central Dogma" of molecular biology.*

Still, Crick's dogma doesn't explain everything about protein construction. For one thing, notice that, with four DNA letters, sixty-four different triplets are possible ($4 \times 4 \times 4 = 64$). Yet all those triplets code for just twenty amino acids in our bodies. Why?

A physicist named George Gamow founded the RNA Tie Club in 1954 in part to figure out this question. A physicist moonlighting in biology might sound odd—Gamow studied radioactivity and Big Bang theory by day—but other carpet-bagging physicists like Richard Feynman joined the club as well. Not only did RNA offer an intellectual challenge, but many physicists felt appalled by their role in creating nuclear bombs. Physics seemed life destroying, biology life restoring. Overall, twenty-four physicists and biologists joined the Tie Club's roster—one for each amino acid, plus four honorary inductees, for each DNA base. Watson and Crick joined (Watson as official club "Optimist," Crick as "Pessimist"), and each member sported a four-dollar bespoke green wool tie with an RNA strand embroidered in gold silk, made by a haberdasher in Los Angeles. Club stationery read, "Do or die, or don't try."

Despite its collective intellectual horsepower, in one way the club ended up looking a little silly historically. Problems of perverse complexity often attract physicists, and certain physics-happy club members (including Crick, a physics Ph.D.) threw

RNA Tie Club members sporting green wool ties with gold silk RNA embroidery. From left, Francis Crick, Alexander Rich, Leslie E. Orgel, James Watson. (Courtesy of Alexander Rich)

themselves into work on DNA and RNA before anyone realized how simple the DNA → RNA → proteins process was. They concentrated especially on how DNA stores its instructions, and for whatever reason they decided early on that DNA must conceal its instructions in an intricate code—a biological cryptogram. Nothing excites a boys' club as much as coded messages, and like ten-year-olds with Cracker Jack decoder rings, Gamow, Crick, and others set out to break this cipher. They were soon scribbling away with pencil and paper at their desks, page after page piling up, their imaginations happily unfettered by doing experiments. They devised solutions clever enough to make Will Shortz smile—"diamond codes" and "triangle codes" and "comma codes" and many forgotten others. These were NSA-ready codes, codes with reversible messages, codes with error-correction mechanisms built in, codes that maximized storage density by using overlapping triplets. The RNA boys especially loved codes that used equivalent anagrams (so CAG = ACG = GCA, etc.). The approach was popular because when they eliminated all the combinatorial redundancies, the number of unique triplets was exactly twenty. In other words, they'd seemingly found a link between twenty and sixty-four—a reason nature just *had to* use twenty amino acids.

In truth, this was so much numerology. Hard biochemical facts soon deflated the code breakers and proved there's no profound reason DNA codes for twenty amino acids and not nineteen or twenty-one. Nor was there any profound reason (as some hoped) that a given triplet called for a given amino acid. The entire system was accidental, something frozen into cells billions of years ago and now too ingrained to replace—the QWERTY keyboard of biology. Moreover, RNA employs no fancy anagrams or error-correcting algorithms, and it doesn't strive to maximize storage space, either. Our code is actually choking on wasteful redundancy: two, four, even six RNA triplets can represent the same amino acid.* A few biocryptographers later admitted feeling annoyed when they compared nature's code to the best of the Tie Club's codes. Evolution didn't seem nearly as clever.

Any disappointment soon faded, however. Solving the DNA/RNA code finally allowed scientists to integrate two separate realms of genetics, gene-as-information and gene-as-chemical, marrying Miescher with Mendel for the first time. And it actually turned out better in some ways that our DNA code is kludgy. Fancy codes have nice features, but the fancier a code gets, the more likely it will break down or sputter. And however crude, our code does one thing well: it keeps life going by minimizing the damage of mutations. It's exactly that talent that Tsutoma Yamaguchi and so many others had to count on in August 1945.

Ill and swooning, Yamaguchi arrived in Nagasaki early on August 8 and staggered home. (His family had assumed him lost; he convinced his wife he wasn't a ghost by showing her his feet, since Japanese ghosts traditionally have none.) Yamaguchi rested that day, swimming in and out of consciousness, but obeyed an order the next day to report to Mitsubishi headquarters in Nagasaki.

He arrived shortly before 11 a.m. Arms and face bandaged, he struggled to relate the magnitude of atomic warfare to his coworkers. But his boss, skeptical, interrupted to browbeat him, dismissing his story as a fable. "You're an engineer," he barked. "Calculate it. How could one bomb…destroy a whole city?" Famous last words. Just as this Nostradamus wrapped up, a white light swelled inside the room. Heat prickled Yamaguchi's skin, and he hit the deck of the ship-engineering office.

"I thought," he later recalled, "the mushroom cloud followed me from Hiroshima."

Eighty thousand people died in Hiroshima, seventy thousand more in Nagasaki. Of the hundreds of thousands of surviving victims, evidence suggests that roughly 150 got caught near both cities on both days, and that a handful got caught within both blast zones, a circle of intense radiation around 1.5 miles wide. Some of these *nijyuu hibakusha*, double-exposure survivors, had stories to make stones weep. (One had burrowed into his wrecked home in Hiroshima, clawed out his wife's blackened bones, and stacked them in a washbasin to return them to her parents in Nagasaki. He was trudging up the street to the parents' house, washbasin under his arm, when the morning air again fell quiet and the sky was once again bleached white.) But of all the reported double victims, the Japanese government has recognized only one official *nijyuu hibakusha*, Tsutomu Yamaguchi.

Shortly after the Nagasaki explosion, Yamaguchi left his shaken boss and office mates and climbed a watchtower on a nearby hill. Beneath another pall of dirty clouds, he watched his cratered-out hometown smolder, including his own house. A tarry radioactive rain began falling, and he struggled down the hill, fearing the worst. But he found his wife, Hisako, and young son, Katsutoshi, safe in an air-raid shelter.

After the exhilaration of seeing them wore off, Yamaguchi felt even more ill than before. In fact, over the next week he did little but lie in the shelter and suffer like Job. His hair fell out. Boils erupted. He vomited incessantly. His face swelled. He lost hearing in one ear. His reburned skin flaked off, and beneath it his flesh glowed raw red "like whale meat" and pierced him with pain. And as badly as Yamaguchi and others suffered during those months, geneticists feared the long-term agony would be equally bad, as mutations slowly began surfacing.

Scientists had known about mutations for a half century by then, but only work on the DNA → RNA → protein process by the Tie Club and others revealed exactly what these mutations consisted of. Most mutations involve typos, the random substitution of a wrong letter during DNA replication: CAG might become CCG, for instance. In "silent" mutations, no harm is done because of the DNA code's redundancy: the before and after triplets call for the same amino acid, so the net effect is like mistyping *grey* for *gray*. But if CAG and CCG lead to different amino acids—a "missense" mutation—the mistake can change a protein's shape and disable it.

Even worse are "nonsense" mutations. When making proteins, cells will continue to translate RNA into amino acids until they encounter one of three "stop" triplets (e.g., UGA), which terminate the process. A nonsense mutation accidentally turns a normal triplet into one of these stop signs, which truncates the protein early and usually disfigures it. (Mutations can also undo stop signs, and the protein runs on and on.) The black mamba of mutations, the "frameshift" mutation, doesn't involve typos. Instead a base disappears, or an extra base squeezes in. Because cells read RNA in consecutive groups of three, an insertion or deletion screws up not only that triplet but every triplet down the line, a cascading catastrophe.

Cells usually correct simple typos right away, but if something goes wrong (and it will), the flub can become permanently fixed in DNA. Every human being alive today was in fact born with dozens of mutations his parents lacked, and a few of those mutations would likely be lethal if we didn't have two copies of every gene, one from each parent, so one can pick up the slack if the other malfunctions. Nevertheless all living organisms continue to accumulate mutations as they age. Smaller creatures that live at high temperatures are especially hard hit: heat on a molecular level is vigorous motion, and the more molecular motion, the more likely something will bump DNA's elbow as it's copying. Mammals are relatively hefty and maintain a constant body temperature, thankfully, but we do fall victim to other mutations. Wherever two T's appear in a row in DNA, ultraviolet sunlight can fuse them together at an odd angle, which kinks DNA. These accidents can kill cells outright or simply irritate them. Virtually all animals (and plants) have special handyman enzymes to fix T-T kinks, but mammals lost them during evolution—which is why mammals sunburn.

Besides spontaneous mutations, outside agents called mutagens can also injure DNA, and few mutagens do more damage than radioactivity. Again, radioactive gamma rays cause free radicals to form, which cleave the phosphate-sugar backbone of DNA. Scientists now know that if just one strand of the double helix snaps, cells can repair the damage easily, often within an hour. Cells have molecular scissors to snip out mangled DNA, and can run enzymes down the track of the undamaged strand

and add the complementary A, C, G, or T at each point. The repair process is quick, simple, and accurate.

Double-strand breaks, though rarer, cause direr problems. Many double breaks resemble hastily amputated limbs, with tattered flaps of single-stranded DNA hanging off both ends. Cells do have two near-twin copies of every chromosome, and if one has a double-strand break, cells can compare its ragged strands to the (hopefully undamaged) other chromosome and perform repairs. But this process is laborious, and if cells sense widespread damage that needs quick repairs, they'll often just slap two hanging flaps together wherever a few bases line up (even if the rest don't), and hastily fill in the missing letters. Wrong guesses here can introduce a dreaded frameshift mutation—and there are plenty of wrong guesses. Cells repairing double-strand breaks get things wrong roughly three thousand times more often than cells simply copying DNA.

Even worse, radioactivity can delete chunks of DNA. Higher creatures have to pack their many coils of DNA into tiny nuclei; in humans, six linear feet cram into a space less than a thousandth of an inch wide. This intense scrunching often leaves DNA looking like a gnarly telephone cord, with the strand crossing itself or folding back over itself many times. If gamma rays happen to streak through and snip the DNA near one of these crossing points, there will be multiple loose ends in close proximity. Cells don't "know" how the original strands lined up (they don't have memories), and in their haste to fix this catastrophe, they sometimes solder together what should be separate strands. This cuts out and effectively deletes the DNA in between.

So what happens after these mutations? Cells overwhelmed with DNA damage can sense trouble and will kill themselves rather than live with malfunctions. This self-sacrifice can spare the body trouble in small doses, but if too many cells die at once, whole organ systems might shut down. Combined with intense burns, these shutdowns led to many deaths in Japan, and some of the victims who didn't die immediately probably wished they had. Survivors remember seeing people's fingernails fall off whole, dropping from their fists like dried shell pasta. They remember human-sized "dolls of charcoal" slumped in alleys. Someone recalled a man struggling along on two stumps, holding a charred baby upside down. Another recalled a shirtless woman whose breasts had burst "like pomegranates."

During his torment in the air-raid shelter in Nagasaki, Yamaguchi—bald, boily, feverish, half deaf—nearly joined this list of casualties. Only dedicated nursing by his family pulled him through. Some of his wounds still required bandages and would for years. But overall he traded Job's life for something like Samson's: his sores mostly healed, his strength returned, his hair grew back. He began working again, first at Mitsubishi, later as a teacher.

Far from escaping unscathed, however, Yamaguchi now faced a more insidious, more patient threat, because if radioactivity doesn't kill cells outright, it can induce mutations that lead to cancer. That link might seem counterintuitive, since mutations generally harm cells, and tumor cells are thriving if anything, growing and dividing at alarming rates. In truth all healthy cells have genes that act like governors on engines, slowing down their rpm's and keeping their metabolisms in check. If a mutation happens to disable a governor, the cell might not sense enough damage to kill itself, but eventually—especially if other genes, like those that control how often a cell divides, also sustain damage—it can start gobbling up resources and choking off neighbors.

Many survivors of Hiroshima and Nagasaki absorbed doses of radiation a hundred times higher—and in one gulp—than the background radiation a normal person absorbs in a year. And the closer survivors got caught to the epicenter, the more deletions and mutations appeared in their DNA. Predictably, cells that divide rapidly spread their DNA damage more quickly, and Japan saw an immediate spike in leukemia, a cancer of prolific white blood cells. The leukemia epidemic started to fade within a decade, but other cancers gained momentum in the meantime— stomach, colon, ovary, lung, bladder, thyroid, breast.

As bad as things were for adults, fetuses proved more vulnerable: any mutation or deletion in utero multiplied over and over in their cells. Many fetuses younger than four weeks spontaneously aborted, and among those that lived, a rash of birth defects, including tiny heads and malformed brains, appeared in late 1945 and early 1946. (The highest measured IQ among the handicapped ones was 68.) And on top of everything else, by the late 1940s, many of the quarter-million *hibakusha* in Japan began to have new children and pass on their exposed DNA.

Experts on radiation could offer little advice on the wisdom of *hibakusha* having children. Despite the high rates of liver or breast or blood cancer, none of the parents' cancerous DNA would get passed to their children, since children inherit only the DNA in sperm and eggs. Sperm or egg DNA could still mutate, of course, perhaps hideously. But no one had actually measured the damage of Hiroshima-like radiation on humans. So scientists had to work on assumptions. Iconoclastic physicist Edward Teller, father of the H-bomb (and RNA Tie Club member), went around suggesting that small pulses of radiation might even benefit humanity—that for all we knew, mutations goosed our genomes. Even among less reckless scientists, not everyone predicted fairy-tale monstrosities and babes with two heads. Hermann Muller had prophesied in the *New York Times* about future generations of Japanese misfortune, but his ideological opposition to Teller and others may have colored his commentary. (In 2011 a toxicologist, after perusing some now-declassified letters between Muller and another colleague, accused them both of lying to the government about the threat that low doses of radioactivity posed for DNA, then manipulating data and later research to cover themselves. Other historians dispute this interpretation.) Even with high doses of radioactivity, Muller ended up backing away from and qualifying his dire early predictions. Most mutations, he reasoned, however harmful, would prove recessive. And the odds of both parents having flaws in the same gene were remote. So at least among the children of survivors, Mom's healthy genes would probably mask any flaws lurking in Dad's, and vice versa.

But again, no one knew anything for certain, and a sword hung over every birth in Hiroshima and Nagasaki for decades, compounding all the normal anxieties of being a parent. This must have been doubly true for Yamaguchi and his wife, Hisako. Both had regained enough vigor by the early 1950s to want more children, no matter the long-term prognosis. And the birth of their first daughter, Naoko, initially supported Muller's reasoning, as she had no visible defects or deformities. Another daughter followed, Toshiko, and she too proved healthy. However bouncing they were at birth, though, both Yamaguchi daughters endured sickly adolescences and adulthoods. They suspect they inherited a genetically compromised immune system from their twice-bombed father and once-bombed mother.

And yet in Japan generally, the long-feared epidemic of cancers and birth defects among children of *hibakusha* never materialized. In fact no large-scale studies have ever found significant evidence that these children had higher rates of any disease, or even higher rates of mutations. Naoko and Toshiko may well have inherited genetic flaws; it's impossible to rule out, and it sure sounds true, intuitively and emotionally. But at least in the vast majority of cases, genetic fallout didn't settle onto the succeeding generation.*

Even many of the people directly exposed to atomic radiation proved more resilient than scientists expected. Yamaguchi's young son Katsutoshi survived for fifty-plus years after Nagasaki before dying of cancer at fifty-eight. Hisako lasted even longer, dying in 2008 of liver and kidney cancer at eighty-eight. The Nagasaki plutonium bomb probably caused both cancers, yes; but at those ages, it's conceivable that either one would have gotten cancer anyway, for unrelated reasons. As for Yamaguchi himself, despite his double exposure at Hiroshima and Nagasaki in 1945, he lived all the way until 2010, sixty-five extra years, finally dying of stomach cancer at ninety-three.

No one can say definitively what set Yamaguchi apart—why he lived so long after being exposed, twice, while others died from a comparative sprinkling of radioactivity. Yamaguchi never underwent genetic testing (at least not extensive testing), and even if he had, medical science might not have known enough to decide. Still, we can hazard educated guesses. First, his cells clearly did a hell of a job repairing DNA, both single-strand breaks and deadly doubles. It's possible he even had slightly superior repair proteins that worked faster or more efficiently, or certain combinations of repair genes that worked especially well together. We can also surmise that although he could hardly have avoided some mutations, they didn't disable key circuits in his cells. Perhaps the mutations fell along stretches of DNA that don't code for proteins. Or perhaps his were mostly "silent" mutations, where the DNA triplet changes but the amino acid, because of redundancy, doesn't. (If that's the case, the kludgy DNA/RNA code that frustrated the Tie Club actually saved him.) Finally, Yamaguchi apparently avoided until late in life any serious damage to the genetic governors in DNA that keep would-be tumors in check. Any or perhaps all of these factors could have spared him.

Or perhaps—and this seems equally likely—he wasn't all that special biologically. Perhaps many others would have survived just as long. And there is, dare I say it, some small hope in that. Even the most deadly weapons ever deployed, weapons that killed tens of thousands at once, that attacked and scrambled people's biological essences, their DNA, didn't wipe out a nation. Nor did they poison the next generation: thousands of children of A-bomb survivors remain alive today, alive and healthy. After more than three billion years of being exposed to cosmic rays and solar radiation and enduring DNA damage in various forms, nature has its safeguards, its methods of mending and preserving the integrity of DNA. And not just the dogmatic DNA whose messages get transcribed into RNA and translated into proteins—but all DNA, including DNA whose subtler linguistic and mathematical patterns scientists are just beginning to explore.*

4

The Musical Score of DNA

What Kinds of Information Does DNA Store?

lthough inadvertent, a certain pun in *Alice's Adventures in Wonderland* has, in recent years, developed a curious resonance with DNA. In real life, *Alice's* author, Lewis Carroll, taught mathematics at Oxford University as Charles Lutwidge Dodgson, and one famous line in *Alice* (famous among geeks, at least) has the Mock Turtle moaning about "different branches of arithmetic—ambition, distraction, uglification, and derision." Just before that eye roller, though, the Mock Turtle says something peculiar. He maintains that during his school days, he studied not reading and writing but "reeling and writhing." It's probably just another groaner, but that last term, *writhing*, has piqued the interest of some mathematically savvy DNA scientists.

Scientists have known for decades that DNA, a long and active molecule, can tangle itself into some horrific snarls. What scientists didn't grasp was why these snarls don't choke our cells. In recent years, biologists have turned to an obscure twig of mathematics called knot theory for answers. Sailors and seamstresses

Lewis Carroll's Mock Turtle cried over memories of studying "reeling and writhing" in school, a complaint that resonates with modern DNA research into knots and tangles. (John Tenniel)

mastered the practical side of knots many millennia ago, and religious traditions as distant as Celtic and Buddhist hold certain knots sacred, but the systematic study of knots began only in the later nineteenth century, in Carroll/Dodgson's Victorian Britain. At that time the polymath William Thomson, Lord Kelvin, proposed that the elements on the periodic table were really microscopic knots of different shapes. For precision's sake, Kelvin defined his atomic knots as closed loops. (Knots with loose ends, somewhat like shoelaces, are "tangles.") And he defined a "unique" knot as a unique pattern of strands crossing over and under each other. So if you can slide the loops around on one knot and jimmy its over-under crossings to make it look like another, they're really the same knot. Kelvin suggested that the unique shape of each knot gave rise to the unique properties of each chemical element. Atomic physicists soon proved this clever theory false, but Kelvin did inspire Scottish physicist P. G. Tait to make a chart of unique knots, and knot theory developed independently from there.

Much of early knot theory involved playing cat's cradle and tallying the results. Somewhat pedantically, knot theorists defined the most trivial knot—O, what laymen call a circle—as the "unknot." They classified other unique knots by the number of over-under crossings and by July 2003 could identify 6,217,553,258 distinct knots with up to twenty-two over-under do-si-dos— roughly one knot per person on earth. Meanwhile other knot theorists had moved beyond taking simple censuses, and devised ways to transform one knot into another. This usually involved snipping the string at an under-over crossing, passing the top strand below, and fusing the snipped ends—which sometimes made knots more complicated but often simplified them. Although studied by legitimate mathematicians, knot theory retained a sense of play throughout. And America's Cup aspirants aside, no one dreamed of applications for knot theory until scientists discovered knotted DNA in 1976.

Knots and tangles form in DNA for a few reasons: its length, its constant activity, and its confinement. Scientists have effectively run simulations of DNA inside a busy nucleus by putting a long, thin rope in a box and jostling it. The rope ends proved quite adept at snaking their way through the rope's coils, and surprisingly complicated knots, with up to eleven crossings, formed in just seconds. (You probably could have guessed this if you've ever dropped earphones into a bag and tried to pull them out later.) Snarls like this can be lethal because the cellular machinery that copies and transcribes DNA needs a smooth track to run on; knots derail it. Unfortunately, the very processes of copying and transcribing DNA can create deadly knots and tangles. Copying DNA requires separating its two strands, but two interlaced helix strands cannot simply be pulled apart, any more than plaits of tightly braided hair can. What's more, when cells do start copying DNA, the long, sticky strings dangling behind sometimes get tangled together. If the strings won't disentangle after a good tug, cells commit suicide — it's that devastating.

Beyond knots per se, DNA can find itself in all sorts of other topological predicaments. Strands can get welded around each other like interlocking links in a chain. They can get twisted excruciatingly tight, like someone wringing out a rag or giving a snake burn on the forearm. They can get wound up into coils tenser than any rattlesnake. And it's this last configuration, the coils, that loops back to Lewis Carroll and his Mock Turtle. Rather imaginatively, knot theorists refer to such coils as "writhes" and refer to the act of coiling as "writhing," as if ropes or DNA were bunched that way in agony. So could the Mock Turtle, per a few recent rumors, have slyly been referring to knot theory with his "reeling and writhing"?

On the one hand, Carroll was working at a prestigious university when Kelvin and Tait began studying knot theory. He might easily have come across their work, and this sort of play math would have appealed to him. Plus, Carroll did write another book called *A Tangled Tale* in which each section — called not chapters but "knots" — consisted of a puzzle to solve. So he certainly incorporated knotty themes into his writing. Still, to be a party pooper, there's good reason to think the Mock Turtle knew nothing about knot theory. Carroll published *Alice* in 1865, some two years before Kelvin broached the idea of knots on the periodic table, at least publicly. What's more, while the term *writhing* might well have been used informally in knot theory before, it first appeared as a technical term in the 1970s. So it seems likely the Mock Turtle didn't progress much past ambition, distraction, uglification, and derision after all.

Nevertheless, even if the punniness of the line postdates Carroll, that doesn't mean we can't enjoy it today. Great literature remains great when it says new things to new generations, and the loops of a knot quite nicely parallel the contours and convolutions of Carroll's plot anyway. What's more, he probably would have delighted at how this whimsical branch of math invaded the real world and became crucial to understanding our biology.

Different combinations of twists and writhes and knots ensure that DNA can form an almost unlimited number of snarls, and what saves our DNA from this torture are mathematically savvy proteins called topoisomerases. Each of these proteins grasps one or two theorems of knot theory and uses them to relieve tension in DNA. Some topoisomerases unlink DNA chains. Other types nick one strand of DNA and rotate it around the other to eliminate twists and writhes. Still others snip DNA wherever it crosses itself, pass the upper strand beneath the lower, and re-fuse them, undoing a knot. Each topoisomerase saves our DNA from a Torquemada-style doom countless time each year, and we couldn't survive without these math nerds. If knot theory sprang from Lord Kelvin's twisted atoms and then went off on its own, it has now circled back to its billions-of-years-old molecular roots in DNA.

Knot theory hasn't been the only unexpected math to pop up during DNA research. Scientists have used Venn diagrams to study DNA, and the Heisenberg uncertainty principle. The architecture of DNA shows traces of the "golden ratio" of length to width found in classical edifices like the Parthenon. Geometry enthusiasts have twisted DNA into Möbius strips and constructed the five Platonic solids. Cell biologists now realize that, to even fit inside the nucleus, long, stringy DNA must fold and refold itself into a fractal pattern of loops within loops within loops, a pattern where it becomes nearly impossible to tell what scale—nano-, micro-, or millimeter—you're looking at. Perhaps most unlikely, in 2011 Japanese scientists used a Tie Club–like code to assign combinations of A, C, G, and T to numbers and letters, then inserted the code for "$E = mc^2$ 1905!" in the DNA of common soil bacteria.

DNA has especially intimate ties to an oddball piece of math called Zipf's law, a phenomenon first discovered by a linguist. George Kingsley Zipf came from solid German stock—his family had run breweries in Germany—and he eventually became a professor of German at Harvard University. Despite his love of language, Zipf didn't believe in owning books, and unlike his colleagues, he lived outside Boston on a seven-acre farm with a vineyard and pigs and chickens, where he chopped down the Zipf family Christmas tree each December. Temperamentally, though, Zipf did not make much of a farmer; he slept through most dawns because he stayed awake most nights studying (from library books) the statistical properties of languages.

A colleague once described Zipf as someone "who would take roses apart to count their petals," and Zipf treated literature no differently. As a young scholar Zipf tackled James Joyce's *Ulysses*, and the main thing he got out of it was that it contained 29,899 different words, and 260,430 words total. From there Zipf dissected *Beowulf*, Homer, Chinese texts, and the oeuvre of the Roman playwright Plautus. By counting the words in each work, he discovered Zipf's law. It says that the most common word in a language appears roughly twice as often as the second most common word, roughly three times as often as the third most common, a hundred times as often as the hundredth most common, and so on. In English, *the* accounts for 7 percent of words, *of* about half that, *and* a third of that, all way down to obscurities like *grawlix* or *boustrophedon*. These distributions hold just as true for Sanskrit, Etruscan, or hieroglyphics as for modern Hindi, Spanish, or Russian. (Zipf also found them in the prices in Sears Roebuck mail-order catalogs.) Even when people make up languages, something like Zipf's law emerges.

After Zipf died in 1950, scholars found evidence of his law in an astonishing variety of other places—in music (more on this later), city population ranks, income distributions, mass extinctions, earthquake magnitudes, the ratios of different colors in paintings and cartoons, and more. Every time, the biggest or most common item in each class was twice as big or common as the second item, three times as big or common as the third, and so on. Probably inevitably, the theory's sudden popularity led to a backlash, especially among linguists, who questioned what Zipf's law even meant, if anything.* Still, many scientists defend Zipf's law because it feels correct—the frequency of words doesn't seem random—and, empirically, it does describe languages in uncannily accurate ways. Even the "language" of DNA.

Of course, it's not apparent at first that DNA is Zipfian, especially to speakers of Western languages. Unlike most languages DNA doesn't have obvious spaces to distinguish each word. It's more like those ancient texts with no breaks or pauses or punctuation of any kind, just relentless strings of letters. You might think that the A-C-G-T triplets that code for amino acids could function as "words," but their individual frequencies don't look Zipfian. To find Zipf, scientists had to look at groups of triplets instead, and a few turned to an unlikely source for help: Chinese search engines. The Chinese language creates compound words by linking adjacent symbols. So if a Chinese text reads ABCD, search engines might examine a sliding "window" to find meaningful chunks, first AB, BC, and CD, then ABC and BCD. Using a sliding window proved a good strategy for finding meaningful chunks in DNA, too. It turns out that, by some measures, DNA looks most Zipfian, most like a language, in groups of around twelve bases. Overall, then, the most mean- ingful unit for DNA might not be a triplet, but four triplets working together—a dodecahedron motif.

The *expression* of DNA, the translation into proteins, also obeys Zipf's law. Like common words, a few genes in every cell get expressed time and time again, while most genes hardly ever come up in conversion. Over the ages cells have learned to rely on these common proteins more and more, and the most com- mon one generally appears twice and thrice and quatrice as often as the next-most-common proteins. To be sure, many scientists harrumph that these Zipfian figures don't mean anything; but others say it's time to appreciate that DNA isn't just analogous to but really functions like a language.

And not just a language: DNA has Zipfian musical proper- ties, too. Given the key of a piece of music, like C major, certain notes appear more often than others. In fact Zipf once investi- gated the prevalence of notes in Mozart, Chopin, Irving Berlin, and Jerome Kern—and lo and behold, he found a Zipfian distri- bution. Later researchers confirmed this finding in other genres, from Rossini to the Ramones, and discovered Zipfian distribu- tions in the timbre, volume, and duration of notes as well.

So if DNA shows Zipfian tendencies, too, is DNA arranged into a musical score of sorts? Musicians have in fact translated the A-C-G-T sequence of serotonin, a brain chemical, into little ditties by assigning the four DNA letters to the notes A, C, G, and, well, E. Other musicians have composed DNA melodies by assigning harmonious notes to the amino acids that popped up most often, and found that this produced more complex and euphonious sounds. This second method reinforces the idea that, much like music, DNA is only partly a strict sequence of "notes." It's also defined by motifs and themes, by how often certain sequences occur and how well they work together. One biologist has even argued that music is a natural medium for studying how genetic bits combine, since humans have a keen ear for how phrases "chunk together" in music.

Something even more interesting happened when two scientists, instead of turning DNA into music, inverted the process and translated the notes from a Chopin nocturne into DNA. They discovered a sequence "strikingly similar" to part of the gene for RNA polymerase. This polymerase, a protein universal throughout life, is what builds RNA from DNA. Which means, if you look closer, that the nocturne actually encodes an entire life cycle. Consider: Polymerase uses DNA to build RNA. RNA in turn builds complicated proteins. These proteins in turn build cells, which in turn build people, like Chopin. He in turn composed harmonious music—which completed the cycle by encoding the DNA to build polymerase. (Musicology recapitulates ontology.)

So was this discovery a fluke? Not entirely. Some scientists argue that when genes first appeared in DNA, they didn't arise randomly, along any old stretch of chromosome. They began instead as repetitive phrases, a dozen or two dozen DNA bases duplicated over and over. These stretches function like a basic musical theme that a composer tweaks and tunes (i.e., mutates) to create pleasing variations on the original. In this sense, then, genes had melody built into them from the start.

Humans have long wanted to link music to deeper, grander themes in nature. Most notably astronomers from ancient Greece right through to Kepler believed that, as the planets ran their course through the heavens, they created an achingly beautiful *musica universalis*, a hymn in praise of Creation. It turns out that universal music does exist, only it's closer than we ever imagined, in our DNA.

Genetics and linguistics have deeper ties beyond Zipf's law. Mendel himself dabbled in linguistics in his older, fatter days, including an attempt to derive a precise mathematical law for how the suffixes of German surnames (like -*mann* and -*bauer*) hybridized with other names and reproduced themselves each generation. (Sounds familiar.) And heck, nowadays, geneticists couldn't even talk about their work without all the terms they've lifted from the study of languages. DNA has synonyms, translations, punctuation, prefixes, and suffixes. Missense mutations (substituting amino acids) and nonsense mutations (interfering with stop codons) are basically typos, while frameshift mutations (screwing up how triplets get read) are old-fashioned typesetting mistakes. Genetics even has grammar and syntax—rules for combining amino acid "words" and clauses into protein "sentences" that cells can read.

More specifically, genetic grammar and syntax outline the rules for how a cell should fold a chain of amino acids into a working protein. (Proteins must be folded into compact shapes before they'll work, and they generally don't work if their shape is wrong.) Proper syntactical and grammatical folding is a crucial part of communicating in the DNA language. However, communication does require more than proper syntax and grammar; a protein sentence has to *mean* something to a cell, too. And, strangely, protein sentences can be syntactically and grammatically perfect, yet have no biological meaning. To understand what on earth that means, it helps to look at something linguist Noam Chomsky once said. He was trying to demonstrate the independence of syntax and meaning in human speech.

His example was "Colorless green ideas sleep furiously." Whatever you think of Chomsky, that sentence has to be one of the most remarkable things ever uttered. It makes no literal sense. Yet because it contains real words, and its syntax and grammar are fine, we can sort of follow along. It's not quite devoid of meaning.

In the same way, DNA mutations can introduce random amino acid words or phrases, and cells will automatically fold the resulting chain together in perfectly syntactical ways based on physics and chemistry. But any wording changes can change the sentence's whole shape and meaning, and whether the result still makes sense depends. Sometimes the new protein sentence contains a mere tweak, minor poetic license that the cell can, with work, parse. Sometimes a change (like a frameshift mutation) garbles a sentence until it reads like grawlix—the #$%^&@! swear words of comics characters. The cell suffers and dies. Every so often, though, the cell reads a protein sentence littered with missense or nonsense...and yet, upon reflection, it somehow does make sense. Something wonderful like Lewis Carroll's "mimsy borogoves" or Edward Lear's "runcible spoon" emerges, wholly unexpectedly. It's a rare beneficial mutation, and at these lucky moments, evolution creeps forward.*

Because of the parallels between DNA and language, scientists can even analyze literary texts and genomic "texts" with the same tools. These tools seem especially promising for analyzing disputed texts, whose authorship or biological origin remains doubtful. With literary disputes, experts traditionally compared a piece to others of known provenance and judged whether its tone and style seemed similar. Scholars also sometimes cataloged and counted what words a text used. Neither approach is wholly satisfactory—the first too subjective, the second too sterile. With DNA, comparing disputed genomes often involves matching up a few dozen key genes and searching for small differences. But this technique fails with wildly different species because the differences are so extensive, and it's not clear which differences are important. By focusing exclusively on genes, this technique also ignores the swaths of regulatory DNA that fall outside genes.

To circumvent these problems, scientists at the University of California at Berkeley invented software in 2009 that again slides "windows" along a string of letters in a text and searches for similarities and patterns. As a test, the scientists analyzed the genomes of mammals and the texts of dozens of books like *Peter Pan*, the Book of Mormon, and Plato's *Republic*. They discovered that the same software could, in one trial run, classify DNA into different genera of mammals, and could also, in another trial run, classify books into different genres of literature with perfect accuracy. In turning to disputed texts, the scientists delved into the contentious world of Shakespeare scholarship, and their software concluded that the Bard did write *The Two Noble Kinsmen*—a play lingering on the margins of acceptance—but didn't write *Pericles*, another doubtful work. The Berkeley team then studied the genomes of viruses and archaebacteria, the oldest and (to us) most alien life-forms. Their analysis revealed new links between these and other microbes and offered new suggestions for classifying them. Because of the sheer amount of data involved, the analysis of genomes can get intensive; the virus-archaebacteria scan monopolized 320 computers for a year. But genome analysis allows scientists to move beyond simple point-by-point comparisons of a few genes and read the full natural history of a species.

Reading a full genomic history, however, requires more dexterity than reading other texts. Reading DNA requires both left-to-right and right-to-left reading—boustrophedon reading.

Otherwise scientists miss crucial palindromes and semordnilaps, phrases that read the same forward and backward (and vice versa).

One of the world's oldest known palindromes is an amazing up-down-and-sideways square carved into walls at Pompeii and other places:

S-A-T-O-R
A-R-E-P-O
T-E-N-E-T
O-P-E-R-A
R-O-T-A-S

At just two millennia old, however, *sator...rotas** falls orders of magnitude short of the age of the truly ancient palindromes in DNA. DNA has even invented two kinds of palindromes. There's the traditional, sex-at-noon-taxes type—GATTA-CATTAG. But because of A-T and C-G base pairing, DNA sports another, subtler type that reads forward down one strand and backward across the other. Consider the string CTAGC-TAG, then imagine what bases must appear on the other strand, GATCGATC. They're perfect palindromes.

Harmless as it seems, this second type of palindrome would send frissons of fear through any microbe. Long ago, many microbes evolved special proteins (called "restriction enzymes") that can snip clean through DNA, like wire cutters. And for what-ever reason, these enzymes can cut DNA only along stretches that are highly symmetrical, like palindromes. Cutting DNA has some useful purposes, like clearing out bases damaged by radiation or relieving tension in knotted DNA. But naughty microbes mostly used these proteins to play Montagues versus Capulets and shred each other's genetic material. As a result microbes have learned the hard way to avoid even modest palindromes.

Not that we higher creatures tolerate many palindromes,

either. Consider CTAGCTAG and GATCGATC again. Notice that the beginning half of either palindromic segment could base-pair with the second half of itself: the first letter with the last (C...G), the second with the penult (T...A), and so on. But for these internal bonds to form, the DNA strand on one side would have to disengage from the other and buckle upward, leaving a bump. This structure, called a "hairpin," can form along any DNA palindrome of decent length because of its inherent symmetry. As you might expect, hairpins can destroy DNA as surely as knots, and for the same reason—they derail cellular machinery.

Palindromes can arise in DNA in two ways. The shortish DNA palindromes that cause hairpins arose randomly, when A's, C's, G's, and T's just happened to arrange themselves symmetrically. Longer palindromes litter our chromosomes as well, and many of those—especially those that wreak havoc on the runt Y chromosome—probably arose through a specific two-step process. For various reasons, chromosomes sometimes accidentally duplicate chunks of DNA, then paste the second copy somewhere down the line. Chromosomes can also (sometimes after double-strand breaks) flip a chunk of DNA by 180 degrees and reattach it ass-backwards. In tandem, a duplication and inversion create a palindrome.

Most chromosomes, though, discourage long palindromes or at least discourage the inversions that create them. Inversions can break up or disable genes, leaving the chromosome ineffective. Inversions can also hurt a chromosome's chances of crossing over—a huge loss. Crossing over (when twin chromosomes cross arms and exchange segments) allows chromosomes to swap genes and acquire better versions, or versions that work better together and make the chromosome more fit. Equally important, chromosomes take advantage of crossing over to perform quality-control checks: they can line up side by side, eyeball each

other up and down, and overwrite mutated genes with nonmu-
tated genes. But a chromosome will cross over only with a partner that looks similar. If the partner looks suspiciously different,
the chromosome fears picking up malignant DNA and refuses
to swap. Inversions look dang suspicious, and in these circumstances, chromosomes with palindromes get shunned.

Y once displayed this intolerance for palindromes. Way back
when, before mammals split from reptiles, X and Y were twin
chromosomes and crossed over frequently. Then, 300 million
years ago, a gene on Y mutated and became a master switch that
causes testes to develop. (Before this, sex was probably determined by the temperature at which Mom incubated her eggs,
the same nongenetic system that determines pink or blue in
turtles and crocodiles.) Because of this change, Y became the
"male" chromosome and, through various processes, accumulated other manly genes, mostly for sperm production. As a consequence, X and Y began to look dissimilar and shied away from
crossing over. Y didn't want to risk its genes being overwritten
by shrewish X, while X didn't want to acquire Y's meathead
genes, which might harm XX females.

After crossing over slowed down, Y grew more tolerant about
inversions, small and large. In fact Y has undergone four massive
inversions in its history, truly huge flips of DNA. Each one created many cool palindromes—one spans three million letters—
but each one made crossing over with X progressively harder.
This wouldn't be a huge deal except, again, crossing over allows
chromosomes to overwrite malignant mutations. Xs could keep
doing this in XX females, but when Y lost its partner, malignant
mutations started to accumulate. And every time one appeared,
cells had no choice but to chop Y down and excise the mutated
DNA. The results were not pretty. Once a large chromosome, Y
has lost all but two dozen of its original fourteen hundred genes.

At that rate, biologists once assumed that Ys were goners. They seem destined to keep picking up dysfunctional mutations and getting shorter and shorter, until evolution did away with Ys entirely—and perhaps did away with males to boot.

Palindromes, however, may have pardoned Y. Hairpins in a DNA strand are bad, but if Y folds *itself* into a giant hairpin, it can bring any two of its palindromes—which are the same genes, one running forward, one backward—into contact. This allows Y to check for mutations and overwrite them. It's like writing down "A man, a plan, a cat, a ham, a yak, a yam, a hat, a canal: Panama!" on a piece of paper, folding the paper over, and correcting any discrepancies letter by letter—something that happens six hundred times in every newborn male. Folding over also allows Ys to make up for their lack of a sex-chromosome partner and "recombine" with themselves, swapping genes at one point along their lengths for genes at another.

This palindromic fix is ingenious. Too clever, in fact, by half. The system Y uses to compare palindromes regrettably doesn't "know" which palindrome has mutated and which hasn't; it just knows there's a difference. So not infrequently, Y overwrites a good gene with a bad one. The self-recombination also tends to—whoops—accidentally delete the DNA between the palindromes. These mistakes rarely kill a man, but can render his sperm impotent. Overall the Y chromosome would disappear if it couldn't correct mutations like this; but the very thing that allows it to, its palindromes, can unman it.

Both the linguistic and mathematical properties of DNA contribute to its ultimate purpose: managing data. Cells store, call up, and transmit messages through DNA and RNA, and scientists routinely speak of nucleic acids encoding and processing information, as if genetics were a branch of cryptography or computer science.

As a matter of fact, modern cryptography has some roots in genetics. After studying at Cornell University, a young geneticist named William Friedman joined an eccentric scientific think tank in rural Illinois in 1915. (It boasted a Dutch windmill, a pet bear named Hamlet, and a lighthouse, despite being 750 miles from the coast.) As Friedman's first assignment, his boss asked him to study the effects of moonlight on wheat genes. But Friedman's statistical background soon got him drawn into another of his boss's lunatic projects*—proving that Francis Bacon not only wrote Shakespeare's plays but left clues throughout the First Folio that trumpeted his authorship. (The clues involved changing the shapes of certain letters.) Although enthused—he'd loved code breaking ever since he'd read Edgar Allan Poe's "The Gold-Bug" as a child—Friedman determined the supposed references to Bacon were bunkum. Someone could use the same deciphering schemes, he noted, to "prove" that Teddy Roosevelt wrote *Julius Caesar*. Nevertheless Friedman had envisioned genetics as biological code breaking, and after his taste of real code breaking, he took a job in cryptography with the U.S. government. Building on the statistical expertise he'd gained in genetics, he soon cracked the secret telegrams that broke the Teapot Dome bribery scandal open in 1923. In the early 1940s he began deciphering Japanese diplomatic codes, including a dozen infamous cables, intercepted on December 6, 1941, from Japan to its embassy in Washington, D.C., that foreshadowed imminent threat.

Friedman had abandoned genetics because genetics in the first decades of the century (at least on farms) involved too much sitting around and waiting for dumb beasts to breed; it was more animal husbandry than data analysis. Had he been born a generation or two later, Friedman would have seen things differently. By the 1950s biologists regularly referred to A-C-G-T base pairs as biological "bits" and to genetics as a "code" to crack. Genetics *became* data analysis, and continued to develop along those lines thanks in part to the work of a younger contemporary of Friedman, an engineer whose work encompassed both cryptography and genetics, Claude Shannon.

Scientists routinely cite Shannon's thesis at MIT, written in 1937 when he was twenty-one years old, as the most important master's thesis ever. In it Shannon outlined a method to combine electronic circuits and elementary logic to do mathematical operations. As a result, he could now design circuits to perform complex calculations—the basis of all digital circuitry. A decade later, Shannon wrote a paper on using digital circuits to encode messages and transmit them more efficiently. It's only barely hyperbole to say that these two discoveries created modern digital communications from scratch.

Amid these seminal discoveries, Shannon indulged his other interests. At the office he loved juggling, and riding unicycles, and juggling while riding unicycles down the hall. At home he tinkered endlessly with junk in his basement; his lifetime inventions include rocket-powered Frisbees, motorized pogo sticks, machines to solve Rubik's Cubes, a mechanical mouse (named Theseus) to solve mazes, a program (named THROBAC) to calculate in Roman numerals, and a cigarette pack–sized "wearable computer" to rip off casinos at roulette.*

Shannon also pursued genetics in his Ph.D. thesis in 1940. At the time, biologists were firming up the connection between genes and natural selection, but the heavy statistics involved frightened many. Though he later admitted he knew squat about genetics, Shannon dived right in and tried to do for genetics what he'd done for electronic circuits: reduce the complexities into simple algebra, so that, given any input (genes in a population), anyone could quickly calculate the output (what genes would thrive or disappear). Shannon spent all of a few months on the paper and, after earning his Ph.D., was seduced by electronics and never got back around to genetics. It didn't matter. His new work became the basis of information theory, a field so widely applicable that it wound its way back to genetics without him.

With information theory, Shannon determined how to transmit messages with as few mistakes as possible — a goal biologists have since realized is equivalent to designing the best genetic code for minimizing mistakes in a cell. Biologists also adopted Shannon's work on efficiency and redundancy in languages. English, Shannon once calculated, was at least 50 percent redundant. (A pulp novel he investigated, by Raymond Chandler, approached 75 percent.) Biologists studied efficiency too because, per natural selection, efficient creatures should be fitter creatures. The less redundancy in DNA, they reasoned, the more information cells could store, and the faster they could process it, a big advantage. But as the Tie Club knew, DNA is sub-suboptimal in this regard. Up to six A-C-G-T triplets code for just one amino acid: totally superfluous redundancy. If cells economized and used fewer triplets per amino acid, they could incorporate more than just the canonical twenty, which would open up new realms of molecular evolution. Scientists have in fact shown that, if coached, cells in the lab can use fifty amino acids.

But if redundancy has costs, it also, as Shannon pointed out, has benefits. A little redundancy in language ensures that we can follow a conversation even if some syllables or words get garbled. Mst ppl hv lttl trbl rdng sntncs wth lttrs mssng. In other words, while too much redundancy wastes time and energy, a little hedges against mistakes. Applied to DNA, we can now see the point of redundancy: it makes a mutation less likely to introduce a wrong amino acid. Furthermore, biologists have calculated that even if a mutation does substitute the wrong amino acid, Mother Nature has rigged things so that, no matter the change, the odds are good that the new amino acid will have similar chemical and physical traits and will therefore get folded properly. It's an amino acid synonym, so cells can still make out the sentence's meaning.

(Redundancy might have uses outside of genes, too. Noncoding DNA—the long expanses of DNA between genes— contains some tediously redundant stretches of letters, places where it looks like someone held his fingers down on nature's keyboard. While these and other stretches look like junk, scientists don't know if they really are expendable. As one scientist mused, "Is the genome a trash novel, from which you can remove a hundred pages and it doesn't matter, or is it more like a Hemingway, where if you remove a page, the story line is lost?" But studies that have applied Shannon's theorems to junk DNA have found that its redundancy looks a lot like the redundancy in languages—which may mean that noncoding DNA has still-undiscovered linguistic properties.)

All of this would have wowed Shannon and Friedman. But perhaps the most fascinating aspect is that, beyond its other clever features, DNA also scooped us on our most powerful information-processing tools. In the 1920s the influential mathematician David Hilbert was trying to determine if there existed any mechanical, turn-the-crank process (an algorithm) that could solve theorems automatically, almost without thought. Hilbert envisioned humans going through this process with pencils and paper. But in 1936 mathematician (and amateur knotologist) Alan Turing sketched out a machine to do the work instead. Turing's machine looked simplistic—just a long recording tape, and a device to move the tape and mark it—but in principle could compute the answer to any solvable problem, no matter how complex, by breaking it down into small, logical steps. The Turing machine inspired many thinkers, Shannon among them. Engineers soon built working models—we call them computers—with long magnetic tapes and recording heads, much as Turing envisioned.

Biologists know, however, that Turing machines resemble nothing so much as the machinery that cells use to copy, mark, and read long strands of DNA and RNA. These Turing biomachines run every living cell, solving all sorts of intricate problems every second. In fact, DNA goes one better than Turing machines: computer hardware still needs software to run; DNA acts as both hardware and software, both storing information and executing commands. It even contains instructions to make more of itself.

And that's not all. If DNA could do only the things we've seen so far—copy itself perfectly over and over, spin out RNA and proteins, withstand the damage of nuclear bombs, encode words and phrases, even whistle a few choice tunes—it would still stand out as an amazing molecule, one of the finest. But what sets DNA apart is its ability to build things billions of times larger than itself—and set them in motion across the globe. DNA has even kept travelogues of everything its creations have seen and done in all that time, and a few lucky creatures can now, finally, after mastering the basics of how DNA works, read these tales for themselves.

PART II

Our Animal Past

Making Things That Crawl and
Frolic and Kill

5

DNA Vindication

Why Did Life Evolve So Slowly —
Then Explode in Complexity?

lmost immediately upon reading the paper, Sister Miriam Michael Stimson must have known that a decade of labor, her life's work, had collapsed. Throughout the 1940s, this Dominican nun—she wore a black-and-white habit (complete with hood) at all times— had carved out a productive, even thriving research career for herself. At small religious colleges in Michigan and Ohio, she had experimented on wound-healing hormones and even helped create a noted hemorrhoid cream (Preparation H) before finding her avocation in studying the shapes of DNA bases.

She progressed quickly in the field and published evidence that DNA bases were changelings—shape-shifters—that could look quite different one moment to the next. The idea was pleasingly simple but had profound consequences for how DNA worked. In 1951, however, two rival scientists obliterated her theory in a single paper, dismissing her work as "slight" and misguided. It was a mortifying moment. As a woman scientist, Sister Miriam carried a heavy burden; she often had to endure patronizing lectures from male colleagues even on her own research topics. And with such a public dismissal, her hard-won reputation was unraveling as quickly and thoroughly as two strands of DNA.

It couldn't have been much consolation to realize, over the next few years, that her repudiation was actually a necessary step in making the most important biological discovery of the century, Watson and Crick's double helix. James Watson and Francis Crick were unusual biologists for their time in that they merely synthesized other people's work and rarely bothered doing experiments. (Even the über-theorist Darwin had run an experimental nursery and trained himself as an expert in barnacles, including barnacle sex.) This habit of "borrowing" got Watson and Crick in trouble sometimes, most notably with Rosalind Franklin, who took the crucial x-rays that illuminated the double helix. But Watson and Crick built on the foundational work of dozens of other, lesser-known scientists as well, Sister Miriam among them. Admittedly, her work wasn't the most important in the field. In fact, her mistakes perpetuated much of the early confusion about DNA. But as with Thomas Hunt Morgan, following along to see how someone faced her errors has its value. And unlike many vanquished scientists, Sister Miriam had the humility, or gumption, to drag herself back into the lab and contribute something to the double helix story in the end.

In many ways biologists in the mid–twentieth century were struggling with the same basic problem—what did DNA look like?—as in Friedrich Miescher's day, when they'd first uncovered its anomalous mix of sugars, phosphates, and ringed bases. Most vexing of all, no one could figure out how the long strands of DNA could possibly mesh and snuggle up together. Today we know that DNA strands mesh automatically because A fits with T, and C with G, but no one knew that in 1950. Everyone assumed letter pairing was random. So scientists had to accommodate every ungainly combination of letters inside their models of DNA: bulky A and G had to fit together sometimes, as did slender C and T. Scientists quickly realized that no matter how these ill-fitting pairs of bases were rotated or jammed together, they would produce dents and bulges, not the sleek DNA shape they expected. At one point Watson and Crick even said to hell with this biomolecular Tetris and wasted months tinkering with an inside-out (and triple-stranded*) DNA model, with the bases facing outward just to get them out of the way.

Sister Miriam tackled an important subproblem of DNA's structure, the precise shapes of the bases. A nun working in a technical field like this might seem strange today, but Miriam recalled later that most women scientists she encountered at conferences and meetings were fellow sisters. Women at the time usually had to relinquish their careers upon marrying, while unmarried women (like Franklin) provoked suspicion or derision and sometimes earned such low pay they couldn't make ends meet. Catholic sisters, meanwhile, respectably unmarried and living in church-run convents, had the financial support and independence to pursue science.

Not that being a sister didn't complicate things, professionally and personally. Like Mendel—born Johann, but Gregor at his monastery—Miriam Stimson and her fellow novices received new names upon entering their convent in Michigan in 1934. Miriam selected Mary, but during the christening ceremony, their archbishop and his assistant skipped an entry on the list, so most women in line were blessed with the wrong handle. No one spoke up, and because no names remained for Miriam, last in line, the clever archbishop used the first name that popped into his head, a man's. The sisterhood was considered a marriage to Christ, and because what God (or archbishops) join together mere humans cannot put asunder, the wrong names became permanent.

These demands for obedience grew more onerous as Sister Miriam started working, and crimped her scientific career. Instead of a full laboratory, her superiors at her small Catholic college spared only a converted bathroom to run experiments in. Not that she had much time to run them: she had to serve as a "wing-nun," responsible for a student dorm, and she had full teaching loads to boot. She also had to wear a habit with an enormous cobra hood even in the lab, which couldn't have made running complicated experiments easy. (She couldn't drive either because the hood obscured her peripheral vision.) Nevertheless Miriam was sincerely clever—friends nicknamed her "M^2"—and as with Mendel, M^2's order promoted and encouraged her love of sci-

Sister Miriam Michael Stimson, a DNA pioneer, wore her enormous hooded habit even in the laboratory. (Archives: Siena Heights University)

ence. Admittedly, they did so partly to combat godless commies in Asia, but partly also to understand God's creation and care for his creatures. Indeed, Miriam and her colleagues contributed to many areas of medicinal chemistry (hence the Preparation H study). The DNA work was a natural extension of this, and she seemed to be making headway in the late 1940s on the shape of DNA bases, by studying their constituent parts.

Carbon, nitrogen, and oxygen atoms make up the core of A, C, G, and T, but the bases also contain hydrogen, which complicates things. Hydrogen hangs out on the periphery of molecules, and as the lightest and most easily peer-pressured element, hydrogen atoms can get yanked around to different positions, giving the bases slightly different shapes. These shifts aren't a big deal—each is pretty much the same molecule before and after—except the position of hydrogen is critical for holding double-stranded DNA together.

Hydrogen atoms consist of one electron circling one proton. But hydrogen usually shares that negative electron with the inner, ringed part of the DNA base. That leaves its positively charged proton derriere exposed. DNA strands bond together by aligning positive hydrogen patches on one strand's bases with negative patches on the other strand's bases. (Negative patches usually center on oxygen and nitrogen, which hoard electrons.) These hydrogen bonds aren't as strong as normal chemical bonds, but that's actually perfect, because cells can unzip DNA when need be.

Though common in nature, hydrogen bonding seemed impossible in DNA in the early 1950s. Hydrogen bonding requires the positive and negative patches to align perfectly—as they do in A and T, and C and G. But again, no one knew that certain letters paired up—and in other letter combinations the charges don't line up so smartly. Research by Sister M^2 and others further fouled up this picture. Her work involved dissolving

DNA bases in solutions with high or low acidity. (High acidity raises the population of hydrogen ions in a solution; low acidity depresses it.) Miriam knew that the dissolved bases and the hydrogen were interacting somehow in her solution: when she shined ultraviolet light through it, the bases absorbed the light differently, a common sign of something changing shape. But she assumed (always risky) that the change involved hydrogens shifting around, and she suggested that this happens naturally in all DNA. If that was true, DNA scientists now had to contemplate hydrogen bonds not only for mismatched bases, but for multiple forms of each mismatched base. Watson and Crick later recalled with exasperation that even textbooks of the time showed bases with hydrogen atoms in different positions, depending on the author's whims and prejudices. This made building models nearly impossible.

As Sister Miriam published papers on this changeling theory of DNA in the late 1940s, she watched her scientific status climb. Pride goeth before the fall. In 1951 two scientists in London determined that acidic and nonacidic solutions did not shift hydrogens around on DNA bases. Instead those solutions either clamped *extra* hydrogens onto them in odd spots or stripped vulnerable hydrogen off. In other words, Miriam's experiments created artificial, nonnatural bases. Her work was useless for determining anything about DNA, and the shape of DNA bases therefore remained enigmatic.

However faulty Miriam's conclusions, though, some experimental techniques she had introduced with this research proved devilishly useful. In 1949 the DNA biologist Erwin Chargaff adapted a method of ultraviolet analysis that Miriam had pioneered. With this technique, Chargaff determined that DNA contains equal amounts of A and T and of C and G. Chargaff never capitalized on the clue, but did blab about it to every scientist he could corner. Chargaff tried to relay this finding to Linus

Pauling—Watson and Crick's main rival—while on a cruise, but Pauling, annoyed at having his holiday interrupted, blew Chargaff off. The cagier Watson and Crick heeded Chargaff (even though he thought them young fools), and from his insight they determined, finally, that A pairs with T, C with G. It was the last clue they needed, and a few degrees of separation from Sister Miriam, the double helix was born.

Except—what about those hydrogen bonds? It's been lost after a half century of hosannas, but Watson and Crick's model rested on an unwarranted, even shaky, assumption. Their bases fit snugly inside the double helix—and fit with proper hydrogen bonds—only if each base had one specific shape, not another. But after Miriam's work was upended, no one knew what shapes the bases had inside living beings.

Determined to help this time, Sister Miriam returned to the lab bench. After the acid-ultraviolet fiasco, she explored DNA with light from the opposite side of the spectrum, the infrared. The standard way to probe a substance with infrared light involved blending it with liquid, but DNA bases wouldn't always mix properly. So Miriam invented a way to mix DNA with a white powder, potassium bromide. To make samples thin enough to study, Miriam's lab team had to borrow a mold from the nearby Chrysler corporation that would shape the powder into "pills" the diameter of an aspirin, then travel to a local machine shop to stamp the pills, with an industrial press, into discs one millimeter thick. The sight of a cab full of nuns in habit descending on a filthy machine shop tickled the grease monkeys on duty, but Miriam remembered them treating her with gentlemanly politeness. And eventually the air force donated a press to her lab so that she could stamp out discs herself. (She had to hold the press down, students remember, long enough to say two Hail Marys.) Because thin layers of potassium bromide were invisible to infrared, the light tickled only the A's, C's, G's, and T's as it streamed through. And over the next decade, infrared studies with the discs (along with other work) proved Watson and Crick right: DNA bases had only one natural shape, the one that produced perfect hydrogen bonds. At this point, and only at this point, could scientists say they grasped DNA's structure.

Of course, understanding its structure wasn't the final goal; scientists had more research ahead. But although M^2 continued to do outstanding work—in 1953 she lectured at the Sorbonne, the first woman scientist to do so since Madame Curie—and although she lived until 2002, reaching the age of eighty-nine, her scientific ambitions petered out along the way. In the swinging 1960s, she doffed her hooded habit for the last time (and learned how to drive), but despite this minor disobedience, she devoted herself to her order during her last decades and stopped experimenting. She let other scientists, including two other women pioneers, unravel how DNA actually builds complex and beautiful life.*

The history of science teems with duplicate discoveries. Natural selection, oxygen, Neptune, sunspots—two, three, even four scientists discovered each independently. Historians continue to bicker about why this happens: perhaps each case was a gigantic coincidence; perhaps one presumed discoverer stole ideas; perhaps the discoveries were impossible before circumstances favored them, and inevitable once they did. But no matter what you believe, scientific simultaneity is a fact. Multiple teams almost sussed out the double helix, and in 1963 two teams did discover another important aspect of DNA. One group was using microscopes to map mitochondria, bean-shaped organs that supply energy inside cells. The other group was pureeing mitochondria and sifting through the guts. Both turned up evidence that mitochondria have their own DNA. In trying to burnish his reputation at the end of the 1800s, Friedrich Miescher had defined the nucleus as the exclusive home for DNA; history once again crossed Friedrich Miescher.

If historical circumstances favor some discoveries, though, science also needs mavericks, those counterclockwise types who see what circumstances blind the rest of us to. Sometimes we even need obnoxious mavericks—because if they're not pugnacious, their theories never penetrate our attention. Such was the case with Lynn Margulis. Most scientists in the mid-1960s explained the origin of mitochondrial DNA rather dully, arguing that cells must have loaned a bit of DNA out once and never gotten it back. But for two decades, beginning in her Ph.D. thesis in 1965, Margulis pushed the idea that mitochondrial DNA was no mere curiosity. She saw it instead as proof of something bigger, proof that life has more ways of mixing and evolving than conventional biologists ever dreamed.

Margulis's theory, endosymbiosis, went like this. We all descended long ago from the first microbes on earth, and all living organisms today share certain genes, on the order of one hundred, as part of that legacy. Soon enough, though, these early microbes began to diverge. Some grew into mammoth blobs, others shrank into specks, and the size difference created opportunities. Most important, some microbes began swallowing and digesting others, while others infected and killed the large and unwary. For either reason, Margulis argued, a large microbe ingested a bug one afternoon long, long ago, and something strange happened: nothing. Either the little Jonah fought off being digested, or his host staved off an internal coup. A standoff followed, and although each one kept fighting, neither could polish the other off. And after untold generations, this initially hostile encounter thawed into a cooperative venture. Gradually the little guy became really good at synthesizing high-octane fuel from oxygen; gradually the whale cell lost its power-producing abilities and specialized instead in providing raw nutrients and shelter. As Adam Smith might have predicted, this division of biolabor benefited each party, and soon neither side could abandon the other without dying. We call the microscopic bugs our mitochondria.

A nice theory overall—but just that. And unfortunately, when Margulis proposed it, scientists didn't respond so nicely. Fifteen journals rejected Margulis's first paper on endosymbiosis, and worse, many scientists outright attacked its speculations. Every time they did, though, she marshaled more evidence and became more pugnacious in emphasizing the independent behavior of mitochondria—that they backstroke about inside cells, that they reproduce on their own schedule, that they have their own cell-like membranes. And their vestigial DNA clinched the case: cells seldom let DNA escape the nucleus to the cellular exurbs, and DNA rarely survives if it tries. We also inherit this DNA differently than chromosome DNA—exclusively from our mothers, since a mother supplies her children with all their mitochondria. Margulis concluded that so-called mtDNA could only have come from once-sovereign cells.

Her opponents countered (correctly) that mitochondria don't work alone; they need chromosomal genes to function, so they're hardly independent. Margulis parried, saying that after three billion years it's not surprising if many of the genes necessary for independent life have faded, until just the Cheshire Cat grin of the old mitochondrial genome remains today. Her opponents didn't buy that—absence of evidence and all—but unlike, say, Miescher, who lacked much backbone for defending himself, Margulis kept swinging back. She lectured and wrote widely on her theory and delighted in rattling audiences. (She once opened a talk by asking, "Any real biologists here? Like molecular biologists?" She counted the raised hands and laughed. "Good. You're going to hate this.")

Biologists did hate endosymbiosis, and the spat dragged on and on until new scanning technology in the 1980s revealed that mitochondria store their DNA not in long, linear chromosomes (as animals and plants do) but in hoops, as bacteria do. The thirty-seven tightly packed genes on the hoop built bacteria-like proteins as well, and the A-C-G-T sequence itself looked remarkably bacterial. Working from this evidence, scientists even identified living relatives of mitochondria, such as typhoid bacteria. Similar work established that chloroplasts—greenish specks that manage photosynthesis inside plants—also contain looped DNA. As with mitochondria, Margulis had surmised that chloroplasts evolved when large ancestor microbes swallowed photosynthesizing pond scum, and Stockholm syndrome ensued. Two cases of endosymbiosis were too much for opponents to explain away. Margulis was vindicated, and she crowed.

In addition to explaining mitochondria, Margulis's theory has since helped solve a profound mystery of life on earth: why evolution damn near stalled after such a promising beginning. Without the kick start of mitochondria, primitive life might never have developed into higher life, much less intelligent human beings.

To see how profound the stall was, consider how easily the universe manufactures life. The first organic molecules on earth probably appeared spontaneously near volcanic vents on the ocean floor. Heat energy there could fuse simple carbon-rich molecules into complex amino acids and even vesicles to serve as crude membranes. Earth also likely imported organics from space. Astronomers have discovered naked amino acids floating in interstellar dust clouds, and chemists have calculated that DNA bases like adenine might form in space, too, since adenine is nothing but five simple HCN molecules (cyanide, of all things) squished into a double ring. Or icy comets might have incubated DNA bases. As ice forms, it gets quite xenophobic and squeezes any organic impurities inside it into concentrated bubbles,

pressure-cooking the gook and making the formation of complicated molecules more likely. Scientists already suspect that
comets filled our seas with water as they bombarded the early
earth, and they might well have seeded our oceans with bio-bits.

From this simmering organic broth, autonomous microorganisms with sophisticated membranes and replaceable moving
parts emerged in just a billion years. (Pretty speedy, if you think
about it.) And from this common beginning, many distinct species popped up in short order, species with distinct livelihoods
and clever ways of carving out a living. After this miracle, however, evolution flatlined: we had many types of truly living creatures, but these microbes didn't evolve much more for well over
a billion years—and might never have.

What doomed them was energy consumption. Primitive
microbes expend 2 percent of their total energy copying and
maintaining DNA, but 75 percent of their energy making proteins from DNA. So even if a microbe develops the DNA for an
advantageous and evolutionarily advanced trait—like an enclosed
nucleus, or a "belly" to digest other microbes, or an apparatus to
communicate with peers—actually building the advanced feature
pretty much depletes it. Adding two is out of the question. In
these circumstances, evolution idles; cells can get only so sophisticated. Cheap mitochondrial power lifted those restrictions. Mitochondria store as much energy per unit size as lightning bolts, and
their mobility allowed our ancestors to add many fancy features at
once, and grow into multifaceted organisms. In fact, mitochondria allowed cells to expand their DNA repertoire 200,000 times,
allowing them not only to invent new genes but to add tons of
regulatory DNA, making them far more flexible when using
genes. This could never have happened without mitochondria,
and we might never have illuminated this evolutionary dark age
without Margulis's theory.

MtDNA opened up whole new realms of science as well, like

genetic archaeology. Because mitochondria reproduce on their own, mtDNA genes are abundant in cells, much more so than chromosome genes. So when scientists go digging around in cavemen or mummies or whatever, they're often rooting out and examining mtDNA. Scientists can also use mtDNA to trace genealogies with unprecedented accuracy. Sperm carry little more than a nuclear DNA payload, so children inherit all their mitochondria from their mothers' much roomier eggs. MtDNA therefore gets passed down largely unchanged through female lines for generation after generation, making it ideal to trace maternal ancestry. What's more, because scientists know how quickly any rare changes do accumulate in a mitochondrial line—one mutation every 3,500 years—they can use mtDNA as a clock: they compare two people's mtDNA, and the more mutations they find, the more years that have passed since the two people shared a maternal ancestor. In fact, this clock tells us that all seven billion people alive today can trace their maternal lineage to one woman who lived in Africa 170,000 years ago, dubbed "Mitochondrial Eve." Eve wasn't the only woman alive then, mind you. She's simply the oldest matrilineal ancestor* of everyone living today.

After mitochondria proved so vital to science, Margulis used her momentum and sudden prestige to advance other offbeat ideas. She began arguing that microbes also donated various locomotor devices to animals, like the tails on sperm, even though those structures lack DNA. And beyond cells merely picking up spare parts, she outlined a grander theory in which endosymbiosis drives all evolution, relegating mutations and natural selection to minor roles. According to this theory, mutations alter creatures only modestly. Real change occurs when genes leap from species to species, or when whole genomes merge, fusing wildly different creatures together. Only after these "horizontal" DNA transfers does natural selection begin, merely to edit out any hopeless monsters that emerge. Meanwhile the hopeful monsters, the beneficiaries of mergers, flourish.

Though Margulis called it revolutionary, in some ways her merger theory just extends a classic debate between biologists who favor (for whatever reasons you want to psychoanalyze) bold leaps and instant species, and biologists who favor conservative adjustments and gradual speciation. The arch-gradualist Darwin saw modest change and common descent as nature's law, and favored a slow-growing tree of life with no overlapping branches. Margulis fell into the radical camp. She argued that mergers can create honest-to-goodness chimeras—blends of creatures technically no different than mermaids, sphinxes, or centaurs. From this point of view, Darwin's quaint tree of life must give way to a quick-spun web of life, with interlocking lines and radials.

However far she strayed into radical ideas, Margulis earned her right to dissent. It's even a little two-faced to praise someone for holding fast to unconventional scientific ideas sometimes and chastise her for not conforming elsewhere; you can't necessarily just switch off the iconoclastic part of the mind when convenient. As renowned biologist John Maynard Smith once admitted, "I think [Margulis is] often wrong, but most of the people I know think it's important to have her around, because she's wrong in such fruitful ways." And lest we forget, Margulis was right in her first big idea, stunningly so. Above all, her work reminds us that pretty plants and things with backbones haven't dominated life's history. Microbes have, and they're the evolutionary fodder from which we multicellular creatures rose.

If Lynn Margulis relished conflict, her older contemporary Barbara McClintock shunned it. McClintock preferred quiet rumination over public confrontation, and her peculiar ideas sprang not from any contrarian streak but from pure flakiness. It's fitting, then, that McClintock dedicated her life to navigating the oddball genetics of plants like maize. By embracing the eccentricity of maize, McClintock expanded our notions of what DNA can do, and provided vital clues for understanding a second great mystery of our evolutionary past: how DNA builds multicellular creatures from Margulis's complex but solo cells.

McClintock's bio divides into two eras: the fulfilled, pre-1951 scientist, and the bitter, post-1951 hermit. Not that things were all daffodils before 1951. From a very young age McClintock squabbled intensely with her mother, a pianist, mostly because Barbara showed more stubborn interest in science and sports like ice skating than the girlish pastimes her mother always said would improve her dating prospects. Her mother even vetoed Barbara's dream to (like Hermann Muller and William Friedman before her) study genetics at Cornell, because nice boys didn't marry brainy gals. Thankfully for science, Barbara's father, a physician, intervened before the fall 1919 semester and packed his daughter off to upstate New York by train.

At Cornell, McClintock flourished, becoming the freshmen women's class president and starring in science classes. Still, her science classmates didn't always appreciate her sharp tongue, especially her lashings over their microscopy work. In that era, preparing microscope samples—slicing cells like deli ham and mounting their gelatinous guts on multiple glass slides without spilling—was intricate and demanding work. Actually using scopes was tricky, too: identifying what specks were what inside a cell could flummox even a good scientist. But McClintock mastered microscopy early, becoming legitimately world-class by graduation. As a graduate student at Cornell, she then honed a technique—"the squash"—that allowed her to flatten whole cells with her thumb and keep them intact on one slide, making them easier to study. Using the squash, she became the first scientist to identify all ten maize chromosomes. (Which, as anyone who's ever gone cross-eyed squinting at the spaghetti mess of chromosomes inside real cells knows, ain't easy.)

Cornell asked McClintock to become a full-time researcher and instructor in 1927, and she began studying how chromosomes interact, with assistance from her best student, Harriet Creighton. Both of these tomboys wore their hair short and often dressed mannishly, in knickers and high socks. People mixed up anecdotes about them, too—like which was it exactly who'd shinnied up a drainpipe one morning after forgetting the keys to her second-story office. Creighton was more outgoing; the reserved McClintock would never have bought a jalopy, as Creighton did, to celebrate the end of World War II and cruise to Mexico. Nonetheless they made a wonderful team, and soon made a seminal discovery. Morgan's fruit fly boys had demonstrated years before that chromosomes probably crossed arms and swapped tips. But their arguments remained statistical, based on abstract patterns. And while many a microscopist had seen chromosomes entangling, no one could tell if they actually exchanged material. But McClintock and Creighton knew every corn chromosome's every knob and carbuncle by sight, and they determined that chromosomes did physically exchange segments. They even linked these exchanges to changes in how genes worked, a crucial confirmation. McClintock dawdled in writing up these results, but when Morgan got wind of them, he insisted she publish, posthaste. She did in 1931. Morgan won his Nobel Prize two years later.

Although pleased with the work—it earned her and Creighton bios in, well, *American Men of Science*—McClintock wanted more. She wanted to study not only chromosomes themselves but how chromosomes changed and mutated, and how those changes built complex organisms with different roots and colors and leaves. Unfortunately, as she tried to set up a lab, social circumstances conspired against her. Like the priesthood, universities at the time offered full professorships only to men (except in home ec), and Cornell had no intention of excepting McClintock. She left reluctantly in 1936 and bounced about, working with Morgan out in California for a spell, then taking research positions in Missouri and Germany. She hated both places.

Truth be told, McClintock had other troubles beyond being the wrong gender. Not exactly the bubbly sort, she'd earned a reputation as sour and uncollegial—she'd once scooped a colleague by tackling his research problem behind his back and publishing her results before he finished. Equally problematic, McClintock worked on corn.

Yes, there was money in corn genetics, since corn was a food crop. (One leading American geneticist, Henry Wallace—future vice president to FDR—made his fortune running a seed company.) Corn also had a scientific pedigree, as Darwin and Mendel had both studied it. Ag scientists even showed an interest in corn mutations: when the United States began exploding nukes at Bikini Atoll in 1946, government scientists placed corn seed beneath the airbursts to study how nuclear fallout affected maize.

McClintock, though, pooh-poohed the traditional ends of corn research, like bigger yields and sweeter kernels. Corn was a means to her, a vehicle to study general inheritance and development. Unfortunately, corn had serious disadvantages for such work. It grew achingly slowly, and its capricious chromosomes often snapped, or grew bulges, or fused, or randomly doubled. McClintock savored the complexity, but most geneticists wanted to avoid these headaches. They trusted McClintock's work—no one matched her on a microscope—but her devotion to corn stranded her between pragmatic scientists helping Iowans grow more bushels and pure geneticists who refused to fuss with unruly corn DNA.

At last, McClintock secured a job in 1941 at rustic Cold

Spring Harbor Laboratory, thirty miles east of Manhattan. Unlike before, she had no students to distract her, and she employed just one assistant—who got a shotgun, and instructions to keep the damn crows off her corn. And although isolated out there with her maize, she was happily isolated. Her few friends always described her as a scientific mystic, constantly chasing the insight that would dissolve the complexity of genetics into unity. "She believed in the great inner lightbulb," one friend remarked. At Cold Spring she had time and space to meditate, and settled into the most productive decade of her career, right through 1951.

Her research actually culminated in March 1950, when a colleague received a letter from McClintock. It ran ten single-spaced pages, but whole paragraphs were scribbled out and scribbled over—not to mention other fervid annotations, connected by arrows and climbing like kudzu up and down the margins. It's the kind of letter you'd think about getting tested for anthrax nowadays, and it described a theory that sounded nutty, too. Morgan had established genes as stationary pearls on a chromosomal necklace. McClintock insisted she'd seen the pearls move—jumping from chromosome to chromosome and burrowing in.

Moreover, these jumping genes somehow affected the color of kernels. McClintock worked with Indian corn, the kind speckled with red and blue and found on harvest floats in parades. She'd seen the jumping genes attack the arms of chromosomes inside these kernels, snapping them and leaving the ends dangling like a compound fracture. Whenever this happened, the kernels stopped producing pigment. Later, though, when the jumping gene got restless and randomly leaped somewhere else, the broken arm healed, and pigment production started up again. Amid her scribbling, McClintock suggested that the break had disrupted the gene for making the pigments.

Indeed, this off/on pattern seemed to explain the randomly colored stripes and swirls of her kernels.

In other words, jumping genes controlled pigment production; McClintock actually called them "controlling elements." (Today they're called transposons or, more generally, mobile DNA.) And like Margulis, McClintock parlayed her fascinating find into a more ambitious theory. Perhaps the knottiest biological question of the 1940s was why cells didn't all look alike: skin and liver and brain cells contain the same DNA, after all, so why didn't they act the same? Previous biologists argued that something in the cell's cytoplasm regulated genes, something external to the nucleus. McClintock had won evidence that chromosomes regulated themselves from within the nucleus—and that this control involved turning genes on or off at the right moments.

In fact, (as McClintock suspected) the ability to turn genes off and on was a crucial step in life's history. After Margulis's complex cells emerged, life once again stalled for over a billion years. Then, around 550 million years ago, huge numbers of multicellular creatures burst into existence. The first beings probably were multicellular by mistake, sticky cells that couldn't free themselves. But over time, by precisely controlling which genes functioned at which moments in which stuck-together cells, the cells could begin to specialize—the hallmark of higher life. Now McClintock thought she had insight into how this profound change came about.

McClintock organized her manic letter into a proper talk, which she delivered at Cold Spring in June 1951. Buoyed by hope, she spoke for over two hours that day, reading thirty-five single-spaced pages. She might have forgiven audience members for nodding off, but to her dismay, she found them merely baffled. It wasn't so much her facts. Scientists knew her reputation, so when she insisted she'd seen genes jump about like fleas, most accepted she had. It was her theory about genetic control that bothered them. Basically, the insertions and jumps seemed too random. This randomness might well explain blue versus red kernels, they granted, but how could jumping genes control all development in multicellular creatures? You can't build a baby or a beanstalk with genes flickering on and off haphazardly. McClintock didn't have good answers, and as the hard questions continued, consensus hardened against her. Her revolutionary idea about controlling elements got downgraded* into another queer property of maize.

Barbara McClintock discovered "jumping genes," but when other scientists questioned her conclusions about them, she became a scientific hermit, crushed and crestfallen. *Inset:* McClintock's beloved maize and microscope. (National Institutes of Health, and Smithsonian Institution, National Museum of American History)

This demotion hurt McClintock badly. Decades after the talk, she still smoldered about colleagues supposedly sniggering at her, or firing off accusations—*You dare question the stationary-gene dogma?* There's little evidence people actually laughed or boiled in rage; again, most accepted jumping genes, just not her theory of control. But McClintock warped the memory into a conspiracy against her. Jumping genes and genetic control had become so interwoven in her heart and mind that attacking one meant attacking both, and attacking her. Crushed, and lacking a brawling disposition, she withdrew from science.*

So began the hermit phase. For three decades, McClintock continued studying maize, often dozing on a cot in her office at night. But she stopped attending conferences and cut off communication with fellow scientists. After finishing experiments, she usually typed up her results as if to submit them to a journal, then filed the paper away without sending it. If her peers dismissed her, she would hurt them back by ignoring them. And in her (now-depressive) solitude, her mystic side emerged fully. She indulged in speculation about ESP, UFOs, and poltergeists, and studied methods of psychically controlling her reflexes. (When visiting the dentist, she told him not to bother with Novocain, as she could lock out pain with her mind.) All the while, she grew maize and squashed slides and wrote up papers that went as unread as Emily Dickinson's poems in her day. She was her own sad scientific community.

Meanwhile something funny was afoot in the larger scientific community, a change almost too subtle to notice at first. The molecular biologists whom McClintock was ignoring began spotting mobile DNA in microbes in the late 1960s. And far from this DNA being a mere novelty, the jumping genes dictated things like whether microbes developed drug resistance. Scientists also found evidence that infectious viruses could (just like mobile DNA) insert genetic material into chromosomes and lurk there permanently. Both were huge medical concerns. Mobile DNA has become vital, too, in tracking evolutionary relationships among species. That's because if you compare a few species, and just two of them have the same transposon burrowed into their DNA at the same point among billions of bases, then those two species almost certainly shared an ancestor recently. More to the point, they shared that ancestor more recently than either shared an ancestor with a third species that lacks the transposon; far too many bases exist for that insertion to have happened twice independently. What look like DNA marginalia, then, actually reveal life's hidden recorded history, and for this and other reasons, McClintock's work suddenly seemed less cute, more profound. As a result her reputation stopped sinking, then rose, year by year. Around 1980 something tipped, and a popular biography of the now-wrinkled McClintock, *A Feeling for the Organism*, appeared in July 1983, making her a minor celebrity. The momentum bucked out of control after that, and unthinkably, just as her own work had done for Morgan a half century before, the adulation propelled McClintock to a Nobel Prize that October.

The hermit had been fairy-tale transformed. She became a latter-day Gregor Mendel, a genius discarded and forgotten — only McClintock lived long enough to see her vindication. Her life soon became a rallying point for feminists and fodder for didactic children's books on never compromising your dreams. That McClintock hated the publicity from the Nobel — it interrupted her research and set reporters prowling about her door — mattered little to fans. And even scientifically, winning the Nobel panged her. The committee had honored her "*discovery* of mobile genetic elements," which was true enough. But in 1951 McClintock had imagined she'd unlocked how genes control other genes and control development in multicellular creatures. Instead scientists honored her, essentially, for her microscope skills—for spotting minnows of DNA darting around. For these reasons, McClintock grew increasingly weary of life post-Nobel, even a little morbid: in her late eighties, she started telling friends she'd surely die at age ninety. Months after her ninetieth birthday party, at James Watson's home, in June 1992, she did indeed pass, cementing her reputation as someone who envisioned things others couldn't.

In the end, McClintock's life's work remained unfulfilled. She did discover jumping genes and vastly expanded our understanding of corn genetics. (One jumping gene, *hopscotch*, seems in fact to have transformed the scrawny wild ancestor of corn into a lush, domesticatable crop in the first place.) More generally McClintock helped establish that chromosomes regulate themselves internally and that on/off patterns of DNA determine a cell's fate. Both ideas remain crucial tenets of genetics. But despite her fondest hopes, jumping genes don't control development or turn genes on and off to the extent she imagined; cells do these things in other ways. In fact it took other scientists many years to explain how DNA accomplishes those tasks—to explain how powerful but isolated cells pulled themselves together long ago and started building truly complex creatures, even creatures as complex as Miriam Michael Stimson, Lynn Margulis, and Barbara McClintock.

6

The Survivors, the Livers

What's Our Most Ancient and Important DNA?

enerations of schoolchildren have learned all about the ruinous amounts of money that European merchants and monarchs spent during colonial days searching for the Northwest Passage—a sailing route to cut horizontally through North America to the spices, porcelain, and tea of Indonesia, India, and Cathay (China). It's less well known that earlier generations of explorers had hunted just as hard for, and believed with no less delusional determination in, a north*east* passage that looped over the frosty top of Russia.

One explorer seeking the northeast passage—Dutchman Willem Barentsz, a navigator and cartographer from the coastal lowlands who was known in English annals as Barents, Barentz, Barentson, and Barentzoon—made his first voyage in 1594 into what's known today as the Barents Sea above Norway. While undertaken for mercenary reasons, voyages like Barentsz's also benefited scientists. Scientific naturalists, while alarmed by the occasional monster that turned up in some savage land, could begin to chart how flora and fauna differ across the globe—work that was a kind of forerunner of our common-descent, common-DNA biology today. Geographers also got much-needed help. Many geographers at the time believed that because of the constant summer sun at high latitudes, polar ice caps melted above a certain point, rendering the North Pole a sunny paradise. And nearly all maps portrayed the pole itself as a monolith of black magnetic rock, which explained why the pole tugged on compasses. In his sally into the Barents Sea, Barentsz aimed to discover if Novaya Zemlya, land north of Siberia, was a promontory of yet another undiscovered continent or merely an island that could be sailed around. He outfitted three ships, *Mercury*, *Swan*, and another *Mercury*, and set out in June 1594.

A few months later, Barentsz and his *Mercury* crew broke with the other ships and began exploring the coast of Novaya Zemlya. In doing so, they made one of the more daring runs in exploration history. For weeks, *Mercury* dodged and ducked a veritable Spanish armada of ice floes, foiling disaster for 1,500 miles. At last Barentsz's men grew frazzled enough to beg to turn back. Barentsz relented, having proved he could navigate the Arctic Sea, and he returned to Holland certain he'd discovered an easy passage to Asia.

Easy, if he avoided monsters. The discovery of the New World and the continued exploration of Africa and Asia had turned up thousands upon thousands of never-dreamed-of plants and animals—and provoked just as many wild tales about beasts that sailors swore they'd seen. For their part, cartographers channeled their inner Hieronymus Bosch and spiced up empty seas and steppes on their maps with wild scenes: blood-red krakens splintering ships, giant otters cannibalizing each other, dragons chewing greedily on rats, trees braining bears with macelike branches, not to mention the ever-popular topless mermaid. One important chart of the era, from 1544, shows a rather contemplative

Monsters of all stripes proved wildly popular on early maps and filled in blank expanses of land and sea for centuries. (Detail from the 1539 *Carta Marina*, a map of Scandinavia, by Olaus Magnus)

Cyclops sitting on the western crook of Africa. Its cartographer, Sebastian Münster, later released an influential compendium of maps interleaved with essays on griffins and avaricious ants that mined gold. Münster also held forth on humanoid-looking beasts around the globe, including the Blemmyae, humans whose faces appeared in their chests; the Cynocephali, people with canine faces; and the Sciopods, grotesque land mermaids with one gargantuan foot, which they used to shade themselves on sunny days by lying down and raising it over their heads. Some of these brutes merely personified (or animalified) ancient fears and superstitions. But amid the farrago of plausible myth and fantastical facts, naturalists could barely keep up.

Even the most scientific naturalist during the age of exploration, Carl von Linné, d.b.a. Linnaeus, speculated on monsters. Linnaeus's *Systema Naturae* set forth the binomial system for naming species that we still use today, inspiring the likes of *Homo sapiens* and *Tyrannosaurus rex*. The book also defined a class of animals called "paradoxa," which included dragons, phoenixes, satyrs, unicorns, geese that sprouted from trees, Heracles's nemesis the Hydra, and remarkable tadpoles that not only got smaller as they aged, but metamorphosed into fish. We might laugh today, but in the last case at least, the joke's on us: shrinking tadpoles do exist, although *Pseudis paradoxa* shrink into regular old frogs, not fish. What's more, modern genetic research reveals a legitimate basis for some of Linnaeus's and Münster's legends.

A few key genes in every embryo play cartographer for other genes and map out our bodies with GPS precision, front to back, left to right, and top to bottom. Insects, fish, mammals, reptiles, and all other animals share many of these genes, especially a subset called *hox* genes. The ubiquity of *hox* in the animal kingdom explains why animals worldwide have the same basic body plan: a cylindrical trunk with a head at one end, an anus at the other, and various appendages sprouting in between. (The Blemmyae, with faces low enough to lick their navels, would be unlikely for this reason alone.)

Unusually for genes, *hox* remain tightly linked after hundreds of millions of years of evolution, almost always appearing together along continuous stretches of DNA. (Invertebrates have one stretch of around ten genes, vertebrates four stretches of basically the same ones.) Even more unusually, each *hox*'s position along that stretch corresponds closely to its assignment in the body. The first *hox* designs the top of the head. The next *hox* designs something slightly lower down. The third *hox* something slightly lower, and so on, until the final *hox* designs our nether regions. Why nature requires this top-to-bottom spatial mapping in *hox* genes isn't known, but again, all animals exhibit this trait.

Scientists refer to DNA that appears in the same basic form

in many, many species as highly "conserved" because creatures remain very careful, very conservative, about changing it. (Some *hox* and *hox*-like genes are so conserved that scientists can rip them out of chickens, mice, and flies and swap them between species, and the genes more or less function the same.) As you might suspect, being highly conserved correlates strongly with the importance of the DNA in question. And it's easy to see, literally see, why creatures don't mess with their highly conserved *hox* genes all that often. Delete one of these genes, and animals can develop multiple jaws. Mutate others and wings disappear, or extra sets of eyes appear in awful places, bulging out on the legs or staring from the ends of antennae. Still other mutations cause genitals or legs to sprout on the head, or cause jaws or antennae to grow in the crotchal region. And these are the lucky mutants; most creatures that gamble with *hox* and related genes don't live to speak of it.

Genes like *hox* don't build animals as much as they instruct other genes on how to build animals: each one regulates dozens of underlings. However important, though, these genes can't control every aspect of development. In particular, they depend on nutrients like vitamin A.

Despite the singular name, vitamin A is actually a few related molecules that we non-biochemists lump together for convenience. These various vitamins A are among the most widespread nutrients in nature. Plants store vitamin A as beta carotene, which gives carrots their distinct color. Animals store vitamin A in our livers, and our bodies convert freely between various forms, which we use in a byzantine array of biochemical processes — to keep eyesight sharp and sperm potent, to boost mitochondria production and euthanize old cells. For these reasons, a lack of vitamin A in the diet is a major health concern worldwide. One of the first genetically enhanced foods created by scientists was so-called golden rice, a cheap source of vitamin A with grains tinted by beta carotene.

Vitamin A interacts with *hox* and related genes to build the fetal brain, lungs, eyes, heart, limbs, and just about every other organ. In fact, vitamin A is so important that cells build special drawbridges in their membranes to let vitamin A, and only vitamin A, through. Once inside a cell, vitamin A binds to special helper molecules, and the resulting complex binds directly to the double helix of DNA, turning *hox* and other genes on. While most signaling chemicals get repulsed at the cell wall and have to shout their instructions through small keyholes, vitamin A gets special treatment, and the *hox* build very little in a baby without a nod from this master nutrient.

But be warned: before you dash to the health store for megadoses of vitamin A for a special pregnant someone, you should know that too much vitamin A can cause substantial birth defects. In fact the body tightly caps its vitamin A concentration, and even has a few genes (like the awkwardly initialed *tgif* gene) that exist largely to degrade vitamin A if its concentration creeps too high. That's partly because high levels of vitamin A in embryos can interfere with the vital, but even more ridiculously named, *sonic hedgehog* gene.

(Yes, it's named for the video game character. A graduate student—one of those wacky fruit fly guys—discovered it in the early 1990s and classified it within a group of genes that, when mutated, cause *Drosophila* to grow spiky quills all over, like hedgehogs. Scientists had already discovered multiple "hedgehog" genes and named them after real hedgehog species, like the Indian hedgehog, moonrat hedgehog, and desert hedgehog. Robert Riddle thought naming his gene after the speedy Sega hero would be funny. By happenstance, *sonic* proved one of the most important genes in the animal repertoire, and the frivolity has not worn well. Flaws can lead to lethal cancers or heartbreaking birth defects, and scientists cringe when they have to explain to some poor family that *sonic hedgehog* will kill a loved one. As one biologist told the *New York Times* about such names, "It's a cute name when you have stupid flies and you call [a gene] *turnip*. When it's linked to development in humans, it's not so cute anymore.")

Just as *hox* genes control our body's top-to-bottom pattern, *shh*—as scientists who detest the name refer to *sonic hedgehog*—helps control the body's left-right symmetry. *Shh* does so by setting up a GPS gradient. When we're still a ball of protoplasm, the incipient spinal column that forms our midline starts to secrete the protein *sonic* produces. Nearby cells absorb lots of it, faraway cells much less. Based on how much protein they absorb, cells "know" exactly where they are in relation to the midline, and therefore know what type of cell they should become.

But if there's too much vitamin A around (or if *shh* fails for a different reason), the gradient doesn't get set up properly. Cells can't figure out their longitude in relation to the midline, and organs start to grow in abnormal, even monstrous ways. In severe cases, the brain doesn't divide into right and left halves; it ends up as one big, undifferentiated blob. The same can happen with the lower limbs: if exposed to too much vitamin A, they fuse together, leading to sirenomelia, or mermaid syndrome. Both fused brains and fused legs are fatal (in the latter case because holes for the anus and bladder don't develop). But the most distressing violations of symmetry appear on the face. Chickens with too much *sonic* have faces with extra-wide midlines, sometimes so wide that two beaks form. (Other animals get two noses.) Too little *sonic* can produce noses with a single giant nostril, or prevent noses from growing at all. In some severe cases, noses appear in the wrong spot, like on the forehead. Perhaps most distressing of all, with too little *sonic*, the two eyes don't start growing where they should, an inch or so to the left and right of the facial midline. Both eyes end up *on* the midline, producing the very Cyclops* that cartographers seemed so silly for including on their maps.

Linnaeus never included Cyclops in his classification scheme, largely because he came to doubt the existence of monsters. He dropped the category of paradoxa from later editions of *Systema Naturae*. But in one case Linnaeus might have been too cynical in dismissing the stories he heard. Linnaeus named the genus that bears reside in, *Ursus*, and he personally named the brown bear *Ursus arctos*, so he knew bears could live in extreme northern climates. Yet he never discussed the existence of polar bears, perhaps because the anecdotes that would have trickled in to him seemed dubious. Who could credit, after all, barroom tales about a ghost-white bear that stalked men across the ice and tore their heads off for sport? Especially when — and people swore this happened, too — if they killed and ate the white bear, it would extract beyond-the-grave revenge by causing their skin to peel off? But such things did happen to Barentsz's men, a horror story that winds its way back and around to the same vitamin A that can produce Cyclopes and mermaids.

Spurred on by the "most exaggerated hopes" of Maurice of Nassau,* a prince of Holland, lords in four Dutch cities filled seven ships with linens, cloths, and tapestries and sent Barentsz back out for Asia in 1595. Haggling delayed the departure until midsummer, and once asea, the captains of the vessels overruled Barentsz (who was merely the navigator) and took a more southerly course than he wished. They did so partly because Barentsz's northerly route seemed mad, and partly because, beyond reaching China, the Dutch seamen were fired by rumors of a remote island whose shores were studded with diamonds. Sure enough, the crew found the island and landed straightaway.

Sailors had been stuffing their pockets with the transparent gems for a number of minutes when, as an olde English account had it, "a great leane white beare came sodainly stealing out" and wrapped his paw around one sailor's neck. Thinking a hirsute sailor had him in a half nelson, he cried out, "Who is that pulles me so by the necke?" His companions, eyes glued to the ground for gems, looked up and just about messed themselves. The polar bear, "falling upon the man, bit his head in sunder, and suckt out his blood."

This encounter opened a centuries-long war between explorers and this "cruell, fierce, and rauenous beast." Polar bears certainly deserved their reputations as mean SOBs. They picked off and devoured any stragglers wherever sailors landed, and they withstood staggering amounts of punishment. Sailors could bury an ax in a bear's back or pump six bullets into its flank—and often, in its rampage, this just made the bear madder. Then again, polar bears had plenty of grievances, too. As one historian notes, "Early explorers seemed to regard it as their duty to kill polar bears," and they piled up carcasses like buffalo hunters later would on the Great Plains. Some explorers deliberately maimed bears to keep as pets and paraded them around in rope nooses. One such bear, hauled aboard a small ship, snapped free from its restraints and, after slapping the sailors about, mutinied and took over the ship. In the bear's fury, though, its noose got tangled in the rudder, and it exhausted itself trying to get free. The brave men retook the ship and butchered the bear.

During the encounter with Barentsz's crew, the bear managed to murder a second sailor, and probably would have kept hunting had reinforcements not arrived from the main ship. A sharpshooter put a bullet clean between the bear's eyes, but the bear shook it off and refused to stop snacking. Other men charged and attacked with swords, but their blades snapped on its head and hide. Finally someone clubbed the beast in the snout and stunned it, enabling another person to slit its throat ear to ear. By this time both sailors had expired, of course, and the rescue squad could do nothing but skin the bear and abandon the corpses.

The rest of the voyage didn't turn out much better for Barentsz's crew. The ships had left too late in the season, and huge floating mines of ice began to threaten their hulls from all sides. The threat swelled each day, and by September, some sailors felt desperate enough to mutiny. Five were hanged. Eventually even Barentsz wilted, fearing the clumsy merchant ships would be stranded in ice. All seven ships returned to port with nothing more than the cargo they'd set out with, and everyone involved lost his frilly shirt. Even the supposed diamonds turned out to be worthless, crumbling glass.

The voyage would have crippled the confidence of a humble man. All Barentsz learned was not to trust superiors. He'd wanted to sail farther north the whole time, so in 1596 he scraped together funds for two more ships and reembarked. Things started smoothly, but once again, Barentsz's ship parted ways with its more prudent companion, helmed by one Captain Rijp. And this time Barentsz pushed too far. He reached the northern tip of Novaya Zemlya and rounded it at last, but as soon as he did so, an unseasonable freeze swept down from the Arctic. The cold stalked his ship southward along the coastline, and each day it became harder to shoulder out room between the floes. Pretty soon Barentsz found himself checkmated, marooned on a continent of ice.

Abandoning their floating coffin—the men could no doubt hear the ice expanding and splintering the ship beneath their feet—the crew staggered ashore to a peninsula on Novaya Zemlya to seek shelter. In their one piece of good luck, they discovered on this treeless island a cache of bleached-white driftwood logs. Naturally the ship's carpenter up and died immediately, but

127

with that wood and a few timbers salvaged from their ship, the dozen crewmen built a log cabin, about eight yards by twelve yards, complete with pine shingles, a porch, and front stairs. With more hope than irony, they called it Het Behouden Huys, the Saved House, and settled down to a grim winter.

Cold was the omnipresent danger, but the Arctic had plenty of other minions to harass the men. In November the sun disappeared for three months, and they grew stir-crazy in their dark, fetid cabin. Perversely, fires threatened them too: the crew almost suffocated from carbon monoxide poisoning one night because of poor ventilation. They managed to shoot some white foxes for fur and meat, but the critters constantly nipped at their food supplies. Even laundry became a black comedy. The men had to stick their clothes almost into their fires to get enough heat to dry them. But the garments could be singed and smoking on one side, and the far side would still be brittle with ice.

For sheer day-to-day terror, however, nothing matched the run-ins with polar bears. One of Barentsz's men, Garrit de Veer, recorded in his diary of the voyage that the bears practically laid siege to Het Behouden Huys, raiding with military precision the barrels of beef, bacon, ham, and fish stacked outside. One bear, smelling a meal on the hearth one night, crept up with such stealth that it had padded up the back stairs and crossed the back door's threshold before anyone noticed. Only a lucky musket shot (which passed through the bear and startled it) prevented a massacre in the tiny quarters.

Fed up, half insane, lusty for revenge, the sailors charged outdoors and followed the blood in the snow until they tracked down and killed the invader. When two more bears attacked over the next two days, the sailors again cut them down. Suddenly in high spirits and hungry for fresh meat, the men decided to feast on the bears, stuffing themselves with anything edible. They gnawed the cartilage off bones and sucked the marrow

Scenes from Barentsz's doomed voyage over the frosty top of Russia. Clockwise from top left: encounters with polar bears; the ship crushed in ice; the hut where the crew endured a grim winter in the 1590s. (Gerrit de Veer, *The Three Voyages of William Barents to the Arctic Regions*)

out, and cooked up all the fleshy victuals—the heart, kidneys, brain, and, most succulent of all, the liver. And with that meal, in a godforsaken cabin at eighty degrees north latitude, European explorers first learned a hard lesson about genetics—a lesson other stubborn Arctic explorers would have to keep learning over and over, a lesson scientists would not understand fully for centuries. Because while polar bear liver may look the same purplish red as any mammal's liver and smell of the same raw ripeness and jiggle the same way on the tines of a fork, there's one big difference: on the molecular level, polar bear liver is supersaturated with vitamin A.

Understanding why that's so awful requires taking a closer look at certain genes, genes that help immature cells in our bodies transform into specialized skin or liver or brain cells or whatever. This was part of the process Barbara McClintock longed to comprehend, but the scientific debate actually predated her by decades.

In the late 1800s, two camps emerged to explain cell specialization, one led by German biologist August Weismann.* Weismann studied zygotes, the fused product of a sperm and egg that formed an animal's first cell. He argued that this first cell obviously contained a complete set of molecular instructions, but that each time the zygote and its daughter cells divided, the cells lost half of those instructions. When cells lost the instructions for all but one type of cell, that's what they became. In contrast, other scientists maintained that cells kept the full set of instructions after each division, but ignored most of the instructions after a certain age. German biologist Hans Spemann decided the issue in 1902 with a salamander zygote. He centered one of these large, soft zygotes in his microscopic crosshairs, waited until it divided into two, then looped a blond strand of his infant daughter Margrette's hair around the boundary between them. (Why Spemann used his daughter's hair isn't clear, since he wasn't bald. Probably the baby's hair was finer.) When he tightened this noose, the two cells split, and Spemann pulled them into different dishes, to develop separately. Weismann would have predicted two deformed half salamanders. But both of Spemann's cells grew into full, healthy adults. In fact, they were genetically identical, which means Spemann had effectively cloned them—in 1902. Scientists had rediscovered Mendel not long before, and Spemann's work implied that cells must retain instructions but turn genes on and off.

Still, neither Spemann nor McClintock nor anyone else could explain how cells turned genes off, the mechanism. That took decades' more work. And it turns out that although cells don't lose genetic information per se, cells do lose access to this information, which amounts to the same thing. We've already seen that DNA must perform incredible acrobatics to fit its entire serpentine length into a tiny cell nucleus. To avoid forming knots during this process, DNA generally wraps itself like a yo-yo string around spools of protein called histones, which then get stacked together and buried inside the nucleus. (Histones were some of the proteins that scientists detected in chromosomes early on, and assumed controlled heredity instead of DNA.) In addition to keeping DNA tangle free, histone spooling prevents cellular machinery from getting at DNA to make RNA, effectively shutting the DNA off. Cells control spooling with chemicals called acetyls. Tacking an acetyl ($COCH_3$) onto a histone unwinds DNA; removing the acetyl flicks the wrist and coils DNA back up.

Cells also bar access to DNA by altering DNA itself, with molecular pushpins called methyl groups (CH_3). Methyls stick best to cytosine, the C in the genetic alphabet, and while methyls don't take up much space—carbon is small, and hydrogen is the smallest element on the periodic table—even that small bump can prevent other molecules from locking onto DNA and turning a gene on. In other words, adding methyl groups mutes genes.

Each of the two hundred types of cells in our bodies has a unique pattern of coiled and methylated DNA, patterns established during our embryonic days. Cells destined to become skin cells must turn off all the genes that produce liver enzymes or neurotransmitters, and something reciprocal happens for everything else. These cells not only remember their pattern for the rest of their lives, they pass on the pattern each time they divide as adult cells. Whenever you hear scientists talking about turning genes on or off, methyls and acetyls are often the culprit. Methyl groups in particular are so important that some scientists have proposed adding a fifth official letter to the DNAlphabet*—A, C, G, T, and now mC, for methylated cytosine.

But for additional and sometimes finer control of DNA, cells turn to "transcription factors" like vitamin A. Vitamin A and other transcription factors bind to DNA and recruit other molecules to start transcribing it. Most important for our purposes, vitamin A stimulates growth and helps convert immature cells into full-fledged bone or muscle or whatever at a fast clip. Vitamin A is especially potent in the various layers of skin. In adults, for instance, vitamin A forces certain skin cells to crawl upward from inside the body to the surface, where they die and become the protective outer layer of skin. High doses of vitamin A can also damage skin through "programmed cell death." This genetic program, a sort of forced suicide, helps the body eliminate sickly cells, so it's not always bad. But for unknown reasons, vitamin A also seems to hijack the system in certain skin cells—as Barentsz's men discovered the hard way.

After the crew tucked into their polar bear stew, rich with burgundy liver chunks, they became more ill than they ever had in their lives. It was a sweaty, fervid, dizzying, bowels-in-a-vice sickness, a real biblical bitch of a plague. In his delirium, the diarist Garrit de Veer remembered the female bear he'd helped butcher, and moaned, "Her death did vs more hurt than her life." Even more distressing, a few days later de Veer realized that many men's skin had begun to peel near their lips or mouths, whatever body parts had touched the liver. De Veer noted with panic that three men fell especially "sicke," and "we verily thought that we should haue lost them, for all their skin came of[f] from the foote to the head."

Only in the mid–twentieth century did scientists determine why polar bear livers contain such astronomical amounts of vitamin A. Polar bears survive mostly by preying on ringed and bearded seals, and these seals raise their young in about the most demanding environment possible, with the 35°F Arctic seas wicking away their body heat relentlessly. Vitamin A enables the seals to survive in this cold: it works like a growth hormone, stimulating cells and allowing seal pups to add thick layers of skin and blubber, and do so quickly. To this end, seal mothers store up whole crates of vitamin A in their livers and draw on this store the whole time they're nursing, to make sure pups ingest enough.

Polar bears also need lots of vitamin A to pack on blubber. But even more important, their bodies tolerate toxic levels of vitamin A because they couldn't eat seals—about the only food source in the Arctic—otherwise. One law of ecology says that poisons accumulate as you move up a food chain, and carnivores at the top ingest the most concentrated doses. This is true of any toxin or any nutrient that becomes toxic at high levels. But unlike many other nutrients, vitamin A doesn't dissolve in water, so when a king predator overdoses, it can't expel the excess through urine. Polar bears either have to deal with all the vitamin A they swallow, or starve. Polar bears adapted by turning their livers into high-tech biohazard containment facilities, to filter vitamin A and keep it away from the rest of the body. (And even with those livers, polar bears have to be careful about intake. They can dine on animals lower on the food chain, with lesser concentrations. But some biologists have wryly noted that if polar bears cannibalized their own livers, they would almost certainly croak.)

Polar bears began evolving their impressive vitamin A–fighting capabilities around 150,000 years ago, when small groups of Alaskan brown bears split off and migrated north to the ice caps. But scientists always suspected that the important genetic changes that made polar bears *polar bears* happened almost right away, instead of gradually over that span. Their reasoning was this. After any two groups of animals split geographically, they begin to acquire different DNA mutations. As the mutations accumulate, the groups develop into different species with different bodies, metabolisms, and behaviors. But not all DNA changes at the same rate in a population. Highly conserved genes like *hox* change grudgingly slowly, at geological paces. Changes in other genes can spread quickly, especially if creatures face environmental stress. For instance, when those brown bears wandered onto the bleak ice sheets atop the Arctic Circle, any beneficial mutations to fight the cold—say, the ability to digest vitamin A–rich seals—would have given some of those bears a substantial boost, and allowed them to have more cubs and take better care of them. And the greater the environmental pressure, the faster such genes can and will spread through a population.

Another way to put this is that DNA clocks—which look at the number and rate of mutations in DNA—tick at different speeds in different parts of the genome. So scientists have to be careful when comparing two species' DNA and dating how long ago they split. If scientists don't take conserved genes or accelerated changes into account, their estimates can be wildly off. With these caveats in mind, scientists determined in 2010 that polar bears had armed themselves with enough cold-weather defenses to become a separate species in as few as twenty thousand years after wandering away from ancestral brown bears— an evolutionary wink.

As we'll see later, humans are Johnny-come-latelies to the meat-eating scene, so it's not surprising that we lack the polar bear's defenses—or that when we cheat up the food chain and eat polar bear livers, we suffer. Different people have different genetic susceptibility to vitamin A poisoning (called hypervitaminosis A), but as little as one ounce of polar bear liver can kill an adult human, and in a ghastly way.

Our bodies metabolize vitamin A to produce retinol, which special enzymes should then break down further. (These enzymes also break down the most common poison we humans ingest, the alcohol in beers, rums, wines, whiskeys, and other booze.)

But polar bear livers overwhelm our poor enzymes with vitamin A, and before they can break it all down, free retinol begins circulating in the blood. That's bad. Cells are surrounded by oil-based membranes, and retinol acts as a detergent and breaks the membranes down. The guts of cells start leaking out incontinently, and inside the skull, this translates to a buildup of fluids that causes headaches, fogginess, and irritability. Retinol damages other tissues as well (it can even crimp straight hair, turning it kinky), but again the skin really suffers. Vitamin A already flips tons of genetic switches in skin cells, causing some to commit suicide, pushing others to the surface prematurely. The burn of more vitamin A kills whole additional swaths, and pretty soon the skin starts coming off in sheets.

We hominids have been learning (and relearning) this same hard lesson about eating carnivore livers for an awfully long time. In the 1980s, anthropologists discovered a 1.6-million-year-old *Homo erectus* skeleton with lesions on his bones characteristic of vitamin A poisoning, from eating that era's top carnivores. After polar bears arose—and after untold centuries of casualties among their people—Eskimos, Siberians, and other northern tribes (not to mention scavenging birds) learned to shun polar bear livers, but European explorers had no such wisdom when they stormed into the Arctic. Many in fact regarded the prohibition on eating livers as "vulgar prejudice," a superstition about on par with worshipping trees. As late as 1900 the English explorer Reginald Koettlitz relished the prospect of digging into polar bear liver, but he quickly discovered that there's sometimes wisdom in taboos. Over a few hours, Koettlitz felt pressure building up inside his skull, until his whole head felt crushed from the inside. Vertigo overtook him, and he vomited repeatedly. Most cruelly, he couldn't sleep it off; lying down made things worse. Another explorer around that time, Dr. Jens Lindhard, fed polar bear liver to nineteen men under his care as an experiment. All became wretchedly ill, so much so that some showed signs of insanity. Meanwhile other starved explorers learned that not just polar bears and seals have toxically elevated levels of vitamin A: the livers of reindeer, sharks, swordfish, foxes, and Arctic huskies* can make excellent last meals as well.

For their part, after being blasted by polar bear liver in 1597, Barentsz's men got wise. As the diarist de Veer had it, after their meal, "there hung a pot still ouer the fire with some of the liuer in it. But the master tooke it and cast it out of the dore, for we had enough of the sawce thereof."

The men soon recovered their strength, but their cabin, clothes, and morale continued to disintegrate in the cold. At last, in June, the ice started melting, and they salvaged rowboats from their ship and headed to sea. They could only dart between small icebergs at first, and they took heavy flak from pursuing polar bears. But on June 20, 1597, the polar ice broke, making real sailing possible. Alas, June 20 also marked the last day on earth of the long-ailing Willem Barentsz, who died at age fifty. The loss of their navigator sapped the courage of the remaining twelve crew members, who still had to cross hundreds of miles of ocean in open boats. But they managed to reach northern Russia, where the locals pitied them with food. A month later they washed ashore on the coast of Lapland, where they ran into, of all people, Captain Jan Corneliszoon Rijp, commander of the very ship that Barentsz had ditched the winter before. Overjoyed—he'd assumed them dead—Rijp carried the men home to the Netherlands* in his ship, where they arrived in threadbare clothes and stunning white fox fur hats.

Their expected heroes' welcome never materialized. That same day another Dutch convoy returned home as well, laden with spices and delicacies from a voyage to Cathay around the southern horn of Africa. Their journey proved that merchant ships could make that long voyage, and while tales of starvation and survival were thrilling, tales of treasure truly stirred the Dutch people's hearts. The Dutch crown granted the Dutch East India Company a monopoly on traveling to Asia via Africa, and an epic trade route was born, a mariner's Silk Road. Barentsz and crew were forgotten.

Perversely, the monopoly on the African route to Asia meant that other maritime entrepreneurs could seek their fortunes only via the northeast passage, so forays into the 500,000-square-mile Barents Sea continued. Finally, salivating over a possible double monopoly, the Dutch East India Company sent its own crew—captained by an Englishman, Henry Hudson—north in 1609. Once again, things got fouled up. Hudson and his ship, the *Half Moon*, crawled north past the tip of Norway as scheduled. But his crew of forty, half of them Dutch, had no doubt heard tales of starvation, exposure, and, God help them, skin peeling off people's bodies head to toe—and mutinied. They forced Hudson to turn west.

If that's what they wanted, Hudson gave it to them, sailing all the way west to North America. He skimmed Nova Scotia and put in at a few places lower down on the Atlantic coast, including a trip up a then-unnamed river, past a skinny swamp island. While disappointed that Hudson had not circled Russia, the Dutch made lemonade by founding a trading colony called New Amsterdam on that isle, Manhattan, within a few years. It's sometimes said of human beings that our passion for exploration is in our genes. With the founding of New York, this was almost literally the case.

7

The Machiavelli Microbe

How Much Human DNA Is Actually Human?

 farmer from Long Island arrived with trepidation at the Rockefeller Institute in Manhattan in 1909, a sick chicken tucked under his arm. A seeming plague of cancer had been wiping out chicken coops across the United States that decade, and the farmer's Plymouth Rock hen had developed a suspicious tumor in its right breast. Nervous about losing his stocks, the farmer turned the hen over to Rockefeller scientist Francis P. Rous, known as Peyton, for a diagnosis. To the farmer's horror, instead of trying to cure the chicken, Rous slaughtered it to get a look at the tumor and run some experiments. Science, though, will be forever grateful for the pollocide.

After extracting the tumor, Rous mashed up a few tenths of an ounce, creating a wet paste. He then filtered it through very small porcelain pores, which removed tumor cells from circulation and left behind mostly just the fluid that exists between cells. Among other things, this fluid helps pass nutrients around, but it can also harbor microbes. Rous injected the fluid into another Plymouth Rock's breast. Soon enough, the second chicken developed a tumor. Rous repeated the experiment with other breeds, like Leghorn fowls, and within six months they developed cancerous masses, about an inch by an inch, too. What was notable, and baffling, about the setup was the filtering step. Since Rous had removed any tumor cells before making the injections, the new tumors couldn't have sprung from old tumor cells reattaching themselves in new birds. The cancer had to have come from the fluid.

Though annoyed, the farmer couldn't have sacrificed his hen to a better candidate to puzzle out the chicken pandemic. Rous, a medical doctor and pathologist, had a strong background with domesticated animals. Rous's father had fled Virginia just before the Civil War and settled in Texas, where he met Peyton's mother. The whole family eventually moved back east to Baltimore, and after high school, Rous enrolled at the Johns Hopkins University, where he supported himself in part by writing, for five dollars a pop, a column called "Wild Flowers of the Month" for the *Baltimore Sun*, about flora in Charm City. Rous gave up the column after entering Johns Hopkins medical school but soon had to suspend his studies. He cut his hand on a tubercular cadaver bone while performing an autopsy, and when he developed tuberculosis himself, school officials ordered a rest cure. But instead of a European-style cure—an actually restful stint in a mountain sanitarium—Rous took a good ol' American cure and worked as a ranch hand in Texas. Though small and slender, Rous loved ranching and developed a keen interest in farm animals. After recovering, he decided to specialize in microbial biology instead of bedside medicine.

All the training on the ranch and in his labs, and all the evidence of the chicken case, pointed Rous to one conclusion. The chickens had a virus, and the virus was spreading cancer. But all his training also told Rous the idea was ridiculous—and his colleagues seconded that. *Contagious cancer, Dr. Rous? How on earth could a virus cause cancer?* Some argued that Rous had misdiagnosed the tumors; perhaps the injections caused an inflammation peculiar to chickens. Rous himself later admitted, "I used to quake in the night for fear that I had made an error." He did publish his results, but even by the circuitous standards of scientific prose, he barely admitted what he believed at some points: "It is perhaps not too much to say that [the discovery] points to the existence of a new group of entities which cause in chickens neoplasms [tumors] of diverse character." But Rous was cagey to be circumspect. A contemporary remembered that his papers on the chicken cancers "were met with reactions ranging from indifference to skepticism and outright hostility."

Most scientists quietly forgot about Rous's work over the next few decades, and for good reason. Because even though a few discoveries from that time linked viruses and cancers biologically, other findings kept pushing them apart. Scientists had determined by the 1950s that cancer cells go haywire in part because their genes malfunction. Scientists had also determined that viruses have small caches of their own genetic material. (Some used DNA; others, like Rous's, used RNA.) Though not technically alive, viruses used that genetic material to hijack cells and make copies of themselves. So viruses and cancer both reproduced uncontrollably, and both used DNA and RNA as common currency—intriguing clues. Meanwhile, however, Francis Crick had published his Central Dogma in 1958, that DNA generates RNA, which generates proteins, in that order. According to this popular understanding of the dogma, RNA viruses like Rous's couldn't possibly disrupt or rewrite the DNA of cells: this would run the dogma backward, which wasn't allowed. So despite the biological overlap, there seemed no way for viral RNA to interface with cancer-causing DNA.

The matter stood at an impasse—data versus dogma—until

a few young scientists in the late 1960s and early 1970s discovered that nature cares little for dogmas. It turns out that certain viruses (HIV is the best-known example) manipulate DNA in heretical ways. Specifically, the viruses can coax an infected cell into reverse-transcribing viral RNA into DNA. Even scarier, they then trick the cell into splicing the newly minted viral DNA back into the cell's genome. In short, these viruses fuse with a cell. They show no respect for the Maginot Line we'd prefer to draw between "their" DNA and "our" DNA.

This strategy for infecting cells might seem convoluted: why would an RNA virus like HIV bother converting itself into DNA, especially when the cell has to transcribe that DNA back into RNA later anyway? It seems even more baffling when you consider how resourceful and nimble RNA is compared to DNA. Lone RNA can build rudimentary proteins, whereas lone DNA sort of just sits there. RNA can also build copies of itself *by itself*, like that M. C. Escher drawing of two hands sketching themselves into existence. For these reasons, most scientists believe that RNA probably predated DNA in the history of life, since early life would have lacked the fancy innards-copying equipment cells have nowadays. (This is called the "RNA world" theory.*)

Nevertheless, early Earth was rough-and-tumble, and RNA is pretty flimsy compared to DNA. Because it's merely single-stranded, RNA letters are constantly exposed to assaults. RNA also has an extra oxygen atom on its ringed sugars, which stupidly cannibalizes its own vertebrae and shreds RNA if it grows too long. So to build anything lasting, anything capable of exploring, swimming, growing, fighting, mating—truly living— fragile RNA had to yield to DNA. This transition to a less corruptible medium a few billion years ago was arguably the most important step in life's history. In a way it resembled the transition in human cultures away from, say, Homer's oral poetry to mute written work: with a stolid DNA text, you miss the RNA-like versatility, the nuances of voice and gesture; but we wouldn't even have the *Iliad* and the *Odyssey* today without papyrus and ink. DNA lasts.

And that's why some viruses convert RNA into DNA after infecting cells: DNA is sturdier, more enduring. Once these *retro*viruses—so named because they run the DNA → RNA → protein dogma backward—weave themselves into a cell's DNA, the cell will faithfully copy the virus genes so long as they both shall live.

The discovery of viral DNA manipulation explained Rous's poor chickens. After the injection, the viruses found their way through the intercellular fluid into the muscle cells. They then ingratiated themselves with the chicken DNA and turned the machinery of each infected muscle cell toward making as many copies of themselves as possible. And it turns out—here's the key—one great strategy for the viruses to spread like mad was to convince the cells harboring viral DNA to spread like mad, too. The viruses did this by disrupting the genetic governors that prevent cells from dividing rapidly. A runaway tumor (and a lot of dead birds) was the result. Transmissible cancers like this are atypical—most cancers have other genetic causes—but for many animals, virus-borne cancers are significant hazards.

This new theory of genetic intrusions was certainly unorthodox. (Even Rous doubted aspects of it.) But if anything, scientists of the time actually underestimated the ability of viruses and other microbes to invade DNA. You can't really attach degrees to the word *ubiquitous;* like being unique, something either is or isn't ubiquitous. But indulge me for a moment in marveling over the complete, total, and utter ubiquity of microbes on a microscopic scale. The little buggers have colonized every known living thing—humans have ten times more microorganisms feasting inside us than we do cells—and saturated every possible ecological niche. There's even a class of viruses that infect only other parasites,* which are scarcely larger than the viruses themselves. For reasons of stability, many of these microbes invade DNA, and they're often nimble enough about changing or masking their DNA to evade and outflank our bodies' defenses. (One biologist calculated that HIV alone has swapped more A's, C's, G's, and T's into and out of its genes in the past few decades than primates have in fifty million years.)

Not until the completion of the Human Genome Project around 2000 did biologists grasp how extensively microbes can infiltrate even higher animals. The name *Human* Genome Project even became something of a misnomer, because it turned out that 8 percent of our genome isn't human at all: a quarter billion of our base pairs are old virus genes. Human genes actually make up less than 2 percent of our total DNA, so by this measure, we're four times more virus than human. One pioneer in studying viral DNA, Robin Weiss, put this evolutionary relationship in stark terms: "If Charles Darwin reappeared today," Weiss mused, "he might be surprised to learn that humans are descended from viruses as well as from apes."

How could this happen? From a virus's point of view, colonizing animal DNA makes sense. For all their deviousness and duplicity, retroviruses that cause cancer or diseases like AIDS are pretty stupid in one respect: they kill their hosts too quickly and die with them. But not all viruses rip through their hosts like microscopic locusts. Less ambitious viruses learn to not disrupt too much, and by showing some restraint they can trick a cell into quietly making copies of themselves for decades. Even better, if viruses infiltrate sperm or egg cells, they can trick their host into passing on viral genes to a new generation, allowing a virus to "live" indefinitely in the host's descendants. (This is happening right now in koalas, as scientists have caught retroviral DNA spreading through koala sperm.) That these viruses have adulterated so much DNA in so many animals hints that this infiltration happens all the time, on scales a little frightening to think about.

Of all the extinct retrovirus genes loaded into human DNA, the vast majority have accumulated a few fatal mutations and no longer function. But otherwise these genes sit intact in our cells and provide enough detail to study the original virus. In fact, in 2006 a virologist in France named Thierry Heidmann used human DNA to resurrect an extinct virus—*Jurassic Park* in a petri dish. It proved startlingly easy. Strings of some ancient viruses appear multiple times in the human genome (the number of copies ranges from dozens to tens of thousands). But the fatal mutations appear, by chance, at different points in each copy. So by comparing many strings, Heidmann could deduce what the original, healthy string must have been simply by counting which DNA letter was most common at each spot. The virus was benign, Heidmann says, but when he reconstituted it and injected it into various mammalian cells—cat, hamster, human—it infected them all.

Rather than wring his hands over this technology (not all ancient viruses are benign) or prophesy doom about it falling into the wrong hands, Heidmann celebrated the resurrection as a scientific triumph, naming his virus *Phoenix* after the mythological bird that rose from its own ashes. Other scientists have replicated Heidmann's work with other viruses, and together they founded a new discipline called paleovirology. Soft, tiny viruses don't leave behind rocky fossils, like the ones paleontologists dig up, but the paleovirologists have something just as informative in fossil DNA.

Close examination of this "wet" fossil record suggests that our genome might be even more than 8 percent virus. In 2009, scientists discovered in humans four stretches of DNA from something called the bornavirus, which has infected hoofed animals since time immemorial. (It's named after a particularly nasty outbreak among horses in 1885, in a cavalry unit near Borna, Germany. Some of the army horses went stark mad and smashed their own skulls in.) About forty million years ago, a few stray bornaviruses jumped into our monkey ancestors and took refuge in their DNA. It had lurked undetected and unsuspected since then because bornavirus is not a retrovirus, so scientists didn't think it had the molecular machinery to convert RNA to DNA and insert itself somewhere. But lab tests prove that bornavirus can indeed somehow weave itself into human DNA in as few as thirty days. And unlike the mute DNA we inherited from retroviruses, two of the four stretches of borna DNA work like bona fide genes.

Scientists haven't pinned down what those genes do, but they might well make proteins we all need to live, perhaps by boosting our immune systems. Allowing a nonlethal virus to invade our DNA probably inhibits other, potentially worse viruses from doing the same. More important, cells can use benign virus proteins to fight off other infections. It's a simple strategy, really: casinos hire card counters, computer security agencies hire hackers, and no one knows how to combat and neutralize viruses better than a reformed germ. Surveys of our genomes suggest that viruses gave us important regulatory DNA as well. For instance, we've long had enzymes in our digestive tracts to break down starches into simpler sugars. But viruses gave us switches to run those same enzymes in our saliva, too. As a result starchy foods taste sweet inside our mouths. We certainly wouldn't have such a starch-tooth for breads, pastas, and grains without these switches.

These cases may be only the beginning. Almost half of human DNA consists of (à la Barbara McClintock) mobile elements and jumping genes. One transposon alone, the 300-base-long *alu*, appears a million times in human chromosomes and forms fully 10 percent of our genome. The ability of this DNA to detach itself from one chromosome, crawl to another, and burrow into it like a tick looks awfully viruslike. You're not supposed to interject feelings into science, but part of the reason it's so fascinating that we're 8 percent (or more) fossilized virus is that it's so creepy that we're 8 percent (or more) fossilized virus. We have an inborn repugnance for disease and impurity, and we see invading germs as something to shun or drive out, not as intimate parts of ourselves—but viruses and viruslike particles have been tinkering with animal DNA since forever. As one scientist who tracked down the human bornavirus genes said, highlighting the singular, "Our whole notion of ourselves as a species is slightly misconceived."

It gets worse. Because of their ubiquity, microbes of all kinds—not just viruses but bacteria and protozoa—can't help but steer animal evolution. Obviously microbes shape a population by killing some creatures through disease, but that's only part of their power. Viruses, bacteria, and protozoa bequeath new genes to animals on occasion, genes that can alter how our bodies work. They can manipulate animal minds as well. One Machiavellian microbe has not only colonized huge numbers of animals without detection, it has stolen animal DNA—and might even use that DNA to brainwash our minds for its own ends.

Sometimes you acquire wisdom the hard way. "You can visualize a hundred cats," Jack Wright once said. "Beyond that, you can't. Two hundred, five hundred, it all looks the same." This wasn't just speculation. Jack learned this because he and his wife, Donna, once owned a Guinness-certified world record 689 housecats.

It started with Midnight. Wright, a housepainter in Ontario, fell in love with a waitress named Donna Belwa around 1970, and they moved in together with Donna's black, long-haired cat. Midnight committed a peccadillo in the yard one night and became pregnant, and the Wrights didn't have the heart to break up her litter. Having more cats around actually brightened the home, and soon after, they felt moved to adopt strays from the local shelter to save them from being put down. Their house became known locally as Cat Crossing, and people began dropping off more strays, two here, five there. When the *National Enquirer* held a contest in the 1980s to determine who had the most cats in one house, the Wrights won with 145. They soon appeared on *The Phil Donahue Show*, and after that, the "donations" really got bad. One person tied kittens to the Wrights' picnic table and drove off; another shipped a cat via courier on a plane—and made the Wrights pay. But the Wrights turned no feline away, even as their brood swelled toward seven hundred.

Bills reportedly ran to $111,000 per year, including individually wrapped Christmas toys. Donna (who began working at home, managing Jack's painting career) rose daily at 5:30 a.m. and spent the next fifteen hours washing soiled cat beds, emptying litter boxes, forcing pills down cats' throats, and adding ice to kitty bowls (the friction of so many cats' tongues made the water too warm to drink otherwise). But above all she spent her days feeding, feeding, feeding. The Wrights popped open 180 tins of cat food each day and bought three extra freezers to fill with pork, ham, and sirloin for the more finicky felines. They eventually took out a second mortgage, and to keep their heavily leveraged bungalow clean, they tacked linoleum to the walls.

Jack and Donna eventually put their four feet down and by the late 1990s had reduced the population of Cat Crossing to just 359. Almost immediately it crept back up, because they couldn't bear to go lower. In fact, if you read between the lines here, the Wrights seemed almost addicted to having cats around— addiction being that curious state of getting acute pleasure and acute anxiety from the same thing. Clearly they loved the cats. Jack defended his cat "family" to the newspapers and gave each cat an individual name,* even the few that refused to leave his closet. At the same time, Donna couldn't hide the torment of being enslaved to cats. "I'll tell you what's hard to eat in here," she once complained, "Kentucky Fried Chicken. Every time I eat it, I have to walk around the house with the plate under my chin." (Partly to keep cats away, partly to deflect cat hair from her sticky drumsticks.) More poignantly, Donna once admitted, "I get a little depressed sometimes. Sometimes I just say, 'Jack, give me a few bucks,' and I go out and have a beer or two. I sit there for a few hours and it's great. It's peaceful—no cats anywhere." Despite these moments of clarity, and despite their mounting distress,* she and Jack couldn't embrace the obvious solution: ditch the damn cats.

To give the Wrights credit, Donna's constant cleaning made their home seem rather livable, especially compared to the prehistoric filth of some hoarders' homes. Animal welfare inspectors not infrequently find decaying cat corpses on the worst premises, even inside a home's walls, where cats presumably burrow to escape. Nor is it uncommon for the floors and walls to rot and suffer structural damage from saturation with cat urine. Most striking of all, many hoarders deny that things are out of control—a classic sign of addiction.

Scientists have only recently begun laying out the chemical and genetic basis of addiction, but growing evidence suggests that cat hoarders cling to their herds at least partly because they're hooked on a parasite, *Toxoplasma gondii*. Toxo is a one-celled protozoan, kin to algae and amoebas; it has eight thousand genes. And though originally a feline pathogen, Toxo has diversified its portfolio and can now infect monkeys, bats, whales, elephants, aardvarks, anteaters, sloths, armadillos, and marsupials, as well as chickens.

Wild bats or aardvarks or whatever ingest Toxo through infected prey or feces, and domesticated animals absorb it indirectly through the feces found in fertilizers. Humans can also absorb Toxo through their diet, and cat owners can contract it through their skin when they handle kitty litter. Overall it infects one-third of people worldwide. When Toxo invades mammals, it usually swims straight for the brain, where it forms tiny cysts, especially in the amygdala, an almond-shaped region in the mammal brain that guides the processing of emotions, including pleasure and anxiety. Scientists don't know why, but the amygdala cysts can slow down reaction times and induce jealous or aggressive behavior in people. Toxo can alter people's sense of smell, too. Some cat hoarders (those most vulnerable to Toxo) become immune to the pungent urine of cats—they stop smelling it. A few hoarders, usually to their great shame, reportedly even crave the odor.

Toxo does even stranger things to rodents, a common meal for cats. Rodents that have been raised in labs for hundreds of generations and have never seen a predator in their whole lives will still quake in fear and scamper to whatever cranny they can find if exposed to cat urine; it's an instinctual, totally hardwired fear. Rats exposed to Toxo have the opposite reaction. They still fear other predators' scents, and they otherwise sleep, mate, navigate mazes, nibble fine cheese, and do everything else normally. But these rats adore cat urine, especially male rats. In fact they more than adore it. At the first whiff of cat urine, their amygdalae throb, as if meeting females in heat, and their testicles swell. Cat urine gets them off.

Toxo toys with mouse desire like this to enrich its own sex life. When living inside the rodent brain, Toxo can split in two and clone itself, the same method by which most microbes reproduce. It reproduces this way in sloths, humans, and other species, too. Unlike most microbes, though, Toxo can also have sex (don't ask) and reproduce sexually—but only in the intestines of cats. It's a weirdly specific fetish, but there it is. Like most organisms, Toxo craves sex, so no matter how many times it has passed its genes on through cloning, it's always scheming to get back inside those erotic cat guts. Urine is its opportunity. By making mice attracted to cat urine, Toxo can lure them toward cats. Cats happily play along, of course, and pounce, and the morsel of mouse ends up exactly where Toxo wanted to be all along, in the cat digestive tract. Scientists suspect that Toxo learned to work its mojo in other potential mammal meals for a similar reason, to ensure that felines of all sizes, from tabbies to tigers, would keep ingesting it.

This might sound like a just-so story so far—a tale that sounds clever but lacks real evidence. Except for one thing. Scientists have discovered that two of Toxo's eight thousand genes help make a chemical called dopamine. And if you know anything about brain chemistry, you're probably sitting up in your chair about now. Dopamine helps activate the brain's reward circuits, flooding us with good feelings, natural highs. Cocaine, Ecstasy, and other drugs also play with dopamine levels, inducing artificial highs. Toxo has the gene for this potent, habit-forming chemical in its repertoire—twice—and whenever an infected brain senses cat urine, consciously or not, Toxo starts pumping it out. As a result, Toxo gains influence over mammalian behavior, and the dopamine hook might provide a plausible biological basis for hoarding cats.*

Toxo isn't the only parasite that can manipulate animals. Much like Toxo, a certain microscopic worm prefers paddling around in the guts of birds but often gets ejected, forcefully, in bird droppings. So the ejected worm wriggles into ants, turns them cherry red, puffs them up like Violet Beauregarde, and convinces other birds they're delicious berries. Carpenter ants also fall victim to a rain-forest fungus that turns them into mindless zombies. First the fungus hijacks an ant's brain, then pilots it toward moist locations, like the undersides of leaves. Upon arriving, the zombified ant bites down, and its jaws lock into place. The fungus turns the ant's guts into a sugary, nutritious goo, shoots a stalk out of its brain, and sends out spores to infect more ants. There's also the so-called Herod bug—the *Wolbachia* bacteria, which infects wasps, mosquitoes, moths, flies, and beetles. *Wolbachia* can reproduce only inside female insects' eggs, so like Herod in the Bible, it often slaughters infant males wholesale, by releasing genetically produced toxins. (In certain lucky insects, *Wolbachia* has mercy and merely fiddles with the genes that determine sex in insects, converting male grubs into females ones—in which case a better nickname might be the Tiresias bug.) Beyond creepy-crawlies, a lab-tweaked version of one virus can turn polygamous male voles—rodents who normally have, as one scientist put it, a "country song… love 'em and leave 'em" attitude toward vole women—into utterly faithful stay-at-home husbands, simply by injecting some repetitive DNA "stutters" into a gene that adjusts brain chemistry. Exposure to the virus arguably even made the voles smarter. Instead of blindly having sex with whatever female wandered near, the males began associating sex with one individual, a trait called "associative learning" that was previously beyond them.

The vole and Toxo cases edge over into uncomfortable territory for a species like us that prizes autonomy and smarts. It's one thing to find broken leftover virus genes in our DNA, quite another to admit that microbes might manipulate our emotions and inner mental life. But Toxo can. Somehow in its long coevolution with mammals, Toxo stole the gene for dopamine production, and the gene has proved pretty successful in influencing animal behavior ever since—both by ramping up pleasure around cats and by damping any natural fear of cats. There's also anecdotal evidence that Toxo can alter other fear signals in the brain, ones unrelated to cats, and convert those impulses into ecstatic pleasure as well. Some emergency room doctors report that motorcycle crash victims often have unusually high numbers of Toxo cysts in their brains. These are the hotshots flying along highways and cutting the S-turns as sharp as possible—the people who get off on risking their lives. And it just so happens that their brains are riddled with Toxo.

It's hard to argue with Toxo scientists who—while thrilled about what Toxo has uncovered about the biology of emotions and the interconnections between fear, attraction, and addiction—also feel creeped out by what their work implies. One Stanford University neuroscientist who studies Toxo says, "It's slightly frightening in some ways. We take fear to be basic and natural. But something can not only eliminate it but turn it into this esteemed thing, attraction. Attraction can be manipulated so as to make us attracted to our worst enemy." That's why Toxo deserves the title of the Machiavellian microbe. Not only can it manipulate us, it can make what's evil seem good.

Peyton Rous's life had a happy if complicated ending. During World War I, he helped establish some of the first blood banks by developing a method to store red blood cells with gelatin and sugar—a sort of blood Jell-O. Rous also bolstered his early work on chickens by studying another obscure but contagious tumor, the giant papilloma warts that once plagued cottontail rabbits in the United States. Rous even had the honor, as editor of a scientific journal, of publishing the first work to firmly link genes and DNA.

Nevertheless, despite this and other work, Rous grew suspicious that geneticists were getting ahead of themselves, and he refused to connect the dots that other scientists were eagerly connecting. For example, before he would publish the paper linking genes and DNA, he made the head scientist strike out a sentence suggesting that DNA was as important to cells as amino acids. Indeed, Rous came to reject the very idea that viruses cause cancer by injecting genetic material, as well as the idea that DNA mutations cause cancer at all. Rous believed viruses promoted cancer in other ways, possibly by releasing toxins; and although no one knows why, he struggled to accept that microbes could influence animal genetics quite as much as his work implied.

Still, Rous never wavered in his conviction that viruses cause tumors somehow, and as his peers unraveled the complicated details of his contagious chicken cancer, they began to appreciate the clarity of his early work all the more. Respect was still grudging in some quarters, and Rous had to endure his much younger son-in-law winning a Nobel Prize for medicine in 1963. But in 1966 the Nobel committee finally vindicated Francis Peyton Rous with a prize of his own. The gap between Rous's important papers and his Nobel, fifty-five years, is one of the longer in prize history. But the win no doubt proved one of the most satisfying, even if he had just four years to enjoy it before dying in 1970. And after his death, it ceased to matter what Rous himself had believed or rejected; young microbiologists, eager to explore how microbes reprogram life, held him up as an idol, and textbooks today cite his work as a classic case of an idea condemned in its own time and later exonerated by DNA evidence.

The story of Cat Crossing also ended in a complicated fashion. As the bills piled higher, creditors nearly seized the Wrights' house. Only donations from cat lovers saved them. Around this time, newspapers also began digging into Jack's past, and reported that, far from being an innocent animal lover, he'd once been convicted of manslaughter for strangling a stripper. (Her body was found on a roof.) Even after these crises passed, the daily hassles continued for Jack and Donna. One visitor reported that "neither had any vacation, any new clothes, any furniture, or draperies." If they rose in the night to go to the bathroom, the dozens of cats on their bed would expand like amoebas to fill the warm hollow, leaving no room to crawl back beneath the covers. "Sometimes you think it'll make you crazy," Donna once confessed. "We can't get away...I cry just about every day in the summertime." Unable to stand the little indignities anymore, Donna eventually moved out. Yet she got drawn back in, unable to walk away from her cats. She returned every dawn to help Jack cope.*

Despite the near certainty of Toxo exposure and infection, no one knows to what extent (if any) Toxo turned Jack and Donna's life inside out. But even if they were infected — and even if neurologists could prove that Toxo manipulated them profoundly — it's hard to censure someone for caring so much for animals. And in a (much, much) larger perspective, the behavior of hoarders might be doing some greater evolutionary good, in the Lynn Margulis sense of mixing up our DNA. Interactions with Toxo and other microbes have certainly influenced our evolution at multiple stages, perhaps profoundly. Retroviruses colonized our genome in waves, and a few scientists argue that it's not a coincidence that these waves appeared just before mammals began to flourish and just before hominid primates emerged. This finding dovetails with another recent theory that microbes may explain Darwin's age-old quandary of the origin of new species. One traditional line of demarcation between species is sexual reproduction: if two populations can't breed and produce viable children, they're separate species. Usually the reproductive barriers are mechanical (animals don't "fit") or biochemical (no viable embryos result). But in one experiment with *Wolbachia* (the Herod-Tiresias bug), scientists took two populations of infected wasps that couldn't produce healthy embryos in the wild and gave them antibiotics. This killed the *Wolbachia* — and suddenly allowed the wasps to reproduce. *Wolbachia* alone had driven them apart.

Along these lines, a few scientists have speculated that if HIV ever reached truly epidemic levels and wiped out most people on earth, then the small percentage of people immune to HIV (and they do exist) could evolve into a new human species. Again, it comes down to sexual barriers. These people couldn't have sex with the nonimmune population (most of us) without killing us off. Any children from the union would have a good chance of dying of HIV, too. And once erected, these sexual and reproductive barriers would slowly but inevitably drive the two populations apart. Even more wildly, HIV, as a retrovirus, could someday insert its DNA into these new humans in a permanent way, joining the genome just as other viruses have. HIV genes would then be copied forever in our descendants, who might have no inkling of the destruction it once wrought.

Of course, saying that microbes infiltrate our DNA may be nothing but a species-centric bias. Viruses have a haiku-like quality about them, some scientists note, a concentration of genetic material that their hosts lack. Some scientists also credit viruses with creating DNA in the first place (from RNA) billions of years ago, and they argue that viruses still invent most new genes today. In fact the scientists who discovered bornavirus DNA in humans think that, far from the bornavirus forcing this DNA into us primates, our chromosomes *stole* this DNA instead. Whenever our mobile DNA starts swimming about, it often grabs other scraps of DNA and drags them along to wherever it's going. Bornavirus replicates only in the nucleus of cells, where our DNA resides, and there's a good chance that mobile DNA mugged the bornavirus long ago, kidnapped its DNA, and kept it around when it proved useful. Along these lines, I've accused Toxo of stealing the dopamine gene from its more sophisticated mammalian hosts. And historical evidence suggests it did. But Toxo also hangs out primarily in the cell nucleus, and there's no theoretical reason we couldn't have stolen this gene from it instead.

It's hard to decide what's less flattering: that microbes out-
smarted our defenses and inserted, wholly by accident, the fancy
genetic tools that mammals needed to make certain evolutionary advances; or that mammals had to shake down little germs
and steal their genes instead. And in some cases these truly were
advances, leaps that helped make us human. Viruses probably
created the mammalian placenta, the interface between mother
and child that allows us to give birth to live young and enables us
to nurture our young. What's more, in addition to producing
dopamine, Toxo can ramp up or ramp down the activity of hundreds of genes inside human neurons, altering how the brain
works. The bornavirus also lives and works between the ears,
and some scientists argue that it could be an important source
for adding variety to the DNA that forms and runs the brain.
This variety is the raw material of evolution, and passing around
microbes like the bornavirus from human to human, probably
via sex, might well have increased the chances of someone getting beneficial DNA. In fact, most microbes responsible for such
boosts likely got passed around via sex. Which means that, if
microbes were as important in pushing evolution forward as
some scientists suggest, STDs could be responsible in some way
for human genius. Descended from apes indeed.

As the virologist Luis Villarreal has noted (and his thoughts
apply to other microbes), "It is our inability to perceive viruses,
especially the silent virus, that has limited our understanding of
the role they play in all life. Only now, in the era of genomics,
can we more clearly see ubiquitous footprints in the genomes of
all life." So perhaps we can finally see too that people who hoard
cats aren't crazy, or at least not merely crazy. They're part of the
fascinating and still-unfolding story of what happens when you
mix animal and microbe DNA.

8

Love and Atavisms

What Genes Make Mammals Mammals?

iven the thousands upon thousands of babies born in and around Tokyo each year, most don't attract much attention, and in December 2005, after forty weeks and five days of pregnancy, a woman named Mayumi quietly gave birth to a baby girl named Emiko. (I've changed the names of family members for privacy's sake.) Mayumi was twenty-eight, and her blood work and sonograms seemed normal throughout her pregnancy. The delivery and its aftermath were also routine—except that for the couple involved, of course, a first child is never routine. Mayumi and her husband, Hideo, who worked at a petrol station, no doubt felt all the normal fluttering anxieties as the ob-gyn cleared the mucus from Emiko's mouth and coaxed her into her first cry. The nurses drew blood from Emiko for routine testing, and again, everything came back normal. They clamped and cut Emiko's umbilical cord, her lifeline to her mother's placenta, and it dried out eventually, and the little stub blackened and fell off in the normal fashion, leaving her with a belly button. A few days later, Hideo and Mayumi left the hospital in Chiba, a sub-urb across the bay from Tokyo, with Emiko in their arms. All perfectly normal.

Thirty-six days after giving birth, Mayumi began bleeding from her vagina. Many women experience vaginal hemorrhages after birth, but three days later, Mayumi also developed a robust fever. With the newborn Emiko to take care of, the couple toughed out Mayumi's spell at home for a few days. But within a week, the bleeding had become uncontrollable, and the family returned to the hospital. Because the wound would not clot, doctors suspected something was wrong with Mayumi's blood. They ordered a round of blood work and waited.

The news was not good. Mayumi tested positive for a grim blood cancer called ALL (acute lymphoblastic leukemia). While most cancers stem from faulty DNA—a cell deletes or miscopies an A, C, G, or T and then turns against the body—Mayumi's cancer had a more complicated origin. Her DNA had undergone what's called a Philadelphia translocation (named after the city in which it was discovered in 1960). A translocation takes place when two *non-twin* chromosomes mistakenly cross over and swap DNA. And unlike a mutational typo, which can occur in any species, this blunder tends to target higher animals with specific genetic features.

Protein-producing DNA—genes—actually makes up very little of the total DNA in higher animals, as little as 1 percent. Morgan's fly boys had assumed that genes almost bumped up against each other on chromosomes, strung along tightly like Alaska's Aleutian Islands. In reality genes are precious rare, scat-tered Micronesian islands in vast chromosomal Pacifics.

So what does all that extra DNA do? Scientists long assumed it did nothing, and snubbed it as "junk DNA." The name has haunted them as an embarrassment ever since. So-called junk DNA actually contains thousands of critical stretches that turn genes on and off or otherwise regulate them—the "junk" manages genes. To take one example, chimpanzees and other primates have short, fingernail-hard bumps (called spines) studding their penises. Humans lack the little prick pricks because sometime in the past few million years, we lost sixty thousand letters of regulatory junk DNA—DNA that would otherwise coax certain genes (which we still have) into making the spines. Besides sparing vaginas, this loss decreases male sensation during sex and thereby prolongs copulation, which scientists suspect helps humans pair-bond and stay monogamous. Other junk DNA fights cancer, or keeps us alive moment to moment.

To their amazement, scientists even found junk DNA—or, as they say now, "noncoding DNA"—cluttering genes themselves. Cells turn DNA into RNA by rote, skipping no letters. But with the full RNA manuscript in hand, cells narrow their eyes, lick a red pencil, and start slashing—think Gordon Lish hacking down Raymond Carver. This editing consists mostly of chopping out unneeded RNA and stitching the remaining bits together to make the actual messenger RNA. (Confusingly, the excised parts are called "introns," the included parts "exons." Leave it to scientists . . .) For example, raw RNA with both exons (capital letters) and introns (lowercase) might read: abcdefGHijklmnOpqrSTuvwxyz. Edited down to exons, it says GHOST.

Lower animals like insects, worms, and their ick ilk contain only a few short introns; otherwise, if introns run on for too long or grow too numerous, their cells get confused and can no longer string something coherent together. The cells of mammals show more aptitude here; we can sift through pages and pages of needless introns and never lose the thread of what the exons are saying. But this talent does have disadvantages. For one, the RNA-editing equipment in mammals must work long, thankless hours: the average human gene contains eight introns, each an average of 3,500 letters long—thirty times longer than the exons they surround. The gene for the largest human protein, *titin*, contains 178 fragments, totaling 80,000 bases, all of which must be stitched together precisely. An even more ridiculously sprawling gene—*dystrophin*, the Jacksonville of human DNA— contains 14,000 bases of coding DNA among 2.2 million bases of intron cruft. Transcription alone takes sixteen hours. Overall this constant splicing wastes incredible energy, and any slipups can ruin important proteins. In one genetic disorder, improper splicing in human skin cells wipes out the grooves and whorls of fingerprints, rendering the fingertips completely bald. (Scientists have nicknamed this condition "immigration delay disease," since these mutants get a lot of guff at border crossings.) Other splicing disruptions are more serious; mistakes in *dystrophin* cause muscular dystrophy.

Animals put up with this waste and danger because introns give our cells versatility. Certain cells can skip exons now and then, or leave part of an intron in place, or otherwise edit the same RNA differently. Having introns and exons therefore gives cells the freedom to experiment: they can produce different RNA at different times or customize proteins for different environments in the body.* For this reason alone, mammals especially have learned to tolerate vast numbers of long introns.

But as Mayumi discovered, tolerance can backfire. Long introns provide places for non-twin chromosomes to get tangled up, since there are no exons to worry about disrupting. The Philadelphia swap takes places along two introns—one on chromosome nine, one on chromosome twenty-two—that are exceptionally long, which raises the odds of these stretches coming into contact. At first our tolerant cells see this swap as no big deal, since it's fiddling "only" with soon-to-be-edited introns. It is a big deal. Mayumi's cells fused two genes that should never be fused—genes that formed, in tandem, a monstrous hybrid protein that couldn't do the job of either individual gene properly. The result was leukemia.

Doctors started Mayumi on chemotherapy at the hospital, but they had caught the cancer late, and she remained quite ill. Worse, as Mayumi deteriorated, their minds started spinning: what about Emiko? ALL is a swift cancer, but not that swift. Mayumi almost certainly had it when pregnant with Emiko. So could the little girl have "caught" the cancer from her mother? Cancer among expectant women is not uncommon, happening once every thousand pregnancies. But none of the doctors had ever seen a fetus catch cancer: the placenta, the organ that connects mother to child, should thwart any such invasion, because in addition to bringing nutrients to the baby and removing waste, the placenta acts as part of the baby's immune system, blocking microbes and rogue cells.

Still, a placenta isn't foolproof—doctors advise pregnant women not to handle kitty litter because Toxo can occasionally slip through the placenta and ravage the fetal brain. And after doing some research and consulting some specialists, the doctors realized that on rare occasions—a few dozen times since the first known instance, in the 1860s—mothers and fetuses came down with cancer simultaneously. No one had ever *proved* anything about the transmission of these cancers, however, because mother, fetus, and placenta are so tightly bound up that questions of cause and effect get tangled up, too. Perhaps the fetus gave the cancer to the mother in these cases. Perhaps they'd both been exposed to unknown carcinogens. Perhaps it was just a sickening coincidence—two strong genetic predispositions for cancer going off at once. But the Chiba doctors, working in 2006, had a tool no previous generation did: genetic sequencing. And as the Mayumi-Emiko case progressed, these doctors used genetic sequencing to pin down, for the first time, whether or not it's possible for a mother to give cancer to her fetus. What's more, their detective work highlighted some functions and mechanisms of DNA unique to mammals, traits that can serve as a springboard for exploring how mammals are genetically special.

Of course, the Chiba doctors weren't imagining their work would take them so far afield. Their immediate concern was treating Mayumi and monitoring Emiko. To their relief, Emiko looked fine. True, she had no idea why her mother had been taken from her, and any breast-feeding—so important to mammalian mothers and children—ceased during chemotherapy. So she certainly felt distress. But otherwise Emiko hit all her growth and development milestones and passed every medical examination. Everything about her seemed, again, normal.

Saying so might creep expectant mothers out, but you can make a good case that fetuses are parasites. After conception, the tiny embryo infiltrates its host (Mom) and implants itself. It proceeds to manipulate her hormones to divert food to itself. It makes Mom ill and cloaks itself from her immune system, which would otherwise destroy it. All well-established games that parasites play. And we haven't even talked about the placenta.

In the animal kingdom, placentas are practically a defining trait of mammals.* Some oddball mammals that split with our lineage long ago (like duck-billed platypi) do lay eggs, just as fish, reptiles, birds, insects, and virtually every other creature do. But of the roughly 2,150 types of mammals, 2,000 have a placenta, including the most widespread and successful mammals, like bats, rodents, and humans. That placental mammals have expanded from modest beginnings into the sea and sky and every other niche from the tropics to the poles suggests that placentas gave them—gave us—a big survival boost.

As probably its biggest benefit, the placenta allows a mammal mother to carry her living, growing children within her. As a result she can keep her children warm inside the womb and run away from danger with them, advantages that spawning-into-water and squatting-on-nest creatures lack. Live fetuses also have longer to gestate and develop energy-intensive organs like the brain; the placenta's ability to pump bodily waste away helps the brain develop, too, since fetuses aren't stewing in toxins. What's more, because she invests so much energy in her developing fetus—not to mention the literal and intimate connection she feels because of the placenta—a mammal mom feels incentive to nurture and watch over her children, sometimes for years. (Or at least feels the need to nag them for years.) The length of this investment is rare among animals, and mammal children reciprocate by forming unusually strong bonds to their mothers. In one sense, then, the placenta, by enabling all this, made us mammals caring creatures.

That makes it all the more creepy that the placenta, in all likelihood, evolved from our old friends the retroviruses. But from a biological standpoint, the connection makes sense. Clamping on to cells happens to be a talent of viruses: they fuse their "envelopes" (their outer skin) to a cell before injecting their genetic material into it. When a ball of embryonic cells swims into the uterus and anchors itself there, the embryo also fuses part of itself with the uterine cells, by using special fusion proteins. And the DNA that primates, mice, and other mammals use to make the fusion proteins appears to be plagiarized from genes that retroviruses use to attach and meld their envelopes. What's more, the uterus of placental mammals draws heavily on other viruslike DNA to do its job, using a special jumping gene called *mer20* to flick 1,500 genes on and off in uterine cells. With both organs, it seems we once again borrowed some handy genetic material from a parasite and adapted it to our own ends. As a bonus, the viral genes in the placenta even provide extra immunity, since the presence of retrovirus proteins (either by warning them off, or outcompeting them) discourages other microbes from circling the placenta.

As another part of its immune function, the placenta filters out any cells that might try to invade the fetus, including cancer cells. Unfortunately, other aspects of the placenta make it downright attractive to cancer. The placenta produces growth hormones to promote the vigorous division of fetal cells, and some cancers thrive on these growth hormones, too. Furthermore, the placenta soaks up enormous amounts of blood and siphons off nutrients for the fetus. That means that blood cancers like leukemia can lurk inside the placenta and flourish. Cancers genetically programmed to metastasize, like the skin cancer melanoma, take to the blood as they slither around inside the body, and they find the placenta quite hospitable as well.

In fact, melanoma is the most common cancer that mothers and fetuses get simultaneously. The first recorded simultaneous cancer, in 1866, in Germany, involved a roaming melanoma that randomly took root in the mother's liver and the child's knee. Both died within nine days. Another horrifying case claimed a twenty-eight-year-old Philadelphia woman, referred to only as "R. McC" by her doctors. It all started when Ms. McC got a brutal sunburn in April 1960. Shortly afterward a half-inch-long mole sprung up between her shoulder blades. It bled whenever she touched it. Doctors removed the mole, and no one thought about it again until May 1963, when she was a few weeks pregnant. During a checkup, doctors noticed a nodule beneath the skin on her stomach. By August the nodule had widened even faster than her belly, and other, painful nodules had sprung up. By January, lesions had spread to her limbs and face, and her doctors opened her up for a cesarean section. The boy inside appeared fine—a full seven pounds, thirteen ounces. But his mother's abdomen was spotted with dozens of tumors, some of them black. Not surprisingly, the birth finished off what little strength she had. Within an hour, her pulse dropped to thirty-six beats per minute, and though her doctors resuscitated her, she died within weeks.

And the McC boy? There was hope at first. Despite the widespread cancer, doctors saw no tumors in Ms. McC's uterus or placenta—her points of contact with her son. And although he was sickly, a careful scan of every crevice and dimple revealed no suspicious-looking moles. But they couldn't check inside him. Eleven days later tiny, dark blue spots began breaking out on the newborn's skin. Things deteriorated quickly after that. The tumors expanded and multiplied, and killed him within seven weeks.

Mayumi had leukemia, not melanoma, but otherwise her family in Chiba reprised the McC drama four decades later. In the hospital, Mayumi's condition deteriorated day by day, her immune system weakened by three weeks of chemotherapy. She finally contracted a bacterial infection and came down with encephalitis, inflammation of the brain. Her body began to convulse and seize—a result of her brain panicking and misfiring—and her heart and lungs faltered, too. Despite intensive care, she died two days after contracting the infection.

Even worse, in October 2006, nine months after burying his wife, Hideo had to return to the hospital with Emiko. The once-bouncing girl had fluid in her lungs and, more troublesome, a raw, fever-red mass disfiguring her right cheek and chin. On an MRI, this premature jowl looked enormous—as large as tiny Emiko's brain. (Try expanding your cheek as far as it will go with air, and its size still wouldn't be close.) Based on its location within the cheek, the Chiba doctors diagnosed sarcoma, cancer of the connective tissue. But with Mayumi in the back of their minds, they consulted experts in Tokyo and England and decided to screen the tumor's DNA to see what they could find.

They found a Philadelphia swap. And not just any Philadelphia swap. Again, this crossover takes place along two tremendously long introns, 68,000 letters long on one chromosome, 200,000 letters long on the other. (This chapter runs about 30,000 letters.) The two arms of the chromosomes could have crossed over at any one of thousands of different points. But the DNA in both Mayumi's and Emiko's cancer had crossed over at the same spot, down the same letter. This wasn't chance. Despite lodging in Emiko's cheek, the cancer basically was the same.

But who gave cancer to whom? Scientists had never solved this mystery before; even the McC case was ambiguous, since the fatal tumors appeared only after the pregnancy started. Doctors pulled out the blood-spot card taken from Emiko at birth and determined that the cancer had been present even then. Further genetic testing revealed that Emiko's normal (non-tumor) cells did *not* show a Philadelphia swap. So Emiko had not inherited any predisposition to this cancer—it had sprung up sometime between conception and the delivery forty weeks later. What's more, Emiko's normal cells also showed, as expected, DNA from both her mother and father. But her cheek tumor cells contained no DNA from Hideo; they were pure Mayumi. This proved, indisputably, that Mayumi had given cancer to Emiko, not vice versa.

Whatever sense of triumph the scientists might have felt, though, was muted. As so often happens in medical research, the most interesting cases spring from the most awful suffering. And in virtually every other historical case where a fetus and mother had cancer simultaneously, both had succumbed to it quickly, normally within a year. Mayumi was already gone, and as the doctors started the eleven-month-old Emiko on chemotherapy, they surely felt these dismal odds weighing on them.

The geneticists on the case felt something different nagging them. The spread of the cancer here was essentially a transplant of cells from one person to another. If Emiko had gotten an organ from her mother or had tissue grafted onto her cheek, her body would have rejected it as foreign. Yet cancer, of all things, had taken root without triggering the placenta's alarms or drawing the wrath of Emiko's immune system. How? Scientists ultimately found the answer in a stretch of DNA far removed from the Philadelphia swap, an area called the MHC.

Biologists back to Linnaeus's time have found it a fascinating exercise to list all the traits that make mammals *mammals*. One place to start—it's the origin of the term, from the Latin for breast, *mamma*—is nursing. In addition to providing nutrition, breast milk activates dozens of genes in suckling infants, mostly in the intestines, but also possibly in spots like the brain. Not to throw any more panic into expectant mothers, but it seems that artificial formula, however similar in carbs, fats, proteins, vitamins, and for all I know taste, just can't goose a baby's DNA the same way.

Other notable traits of mammals include our hair (even whales and dolphins have a comb-over), our unique inner ear and jaw structure, and our odd habit of chewing food before swallowing (reptiles have no such manners). But on a microscopic level, one place to hunt for the origin of mammals is the MHC, the major histocompatibility complex. Nearly all vertebrates have an MHC, a set of genes that helps the immune system. But the MHC is particularly dear to mammals. It's among the most gene-rich stretches of DNA we have, over one hundred genes packed into a small area. And similar to our intron/exon editing equipment, we have more sophisticated and more extensive MHCs than other creatures.* Some of those hundred genes have over a thousand different varieties in humans, offering a virtually unlimited number of combinations to inherit. Even close relatives can differ substantially in their MHC, and the differences among random people are a hundred times higher than those along most other stretches of DNA. Scientists sometimes say that humans are over 99 percent genetically identical. Not along their MHCs they aren't.

MHC proteins basically do two things. First, some of them grab a random sampling of molecules from inside a cell and put them on display on the cellular surface. This display lets other cells, especially "executioner" immune cells, know what's going on inside the cell. If the executioner sees the MHC mounting nothing but normal molecules, it ignores the cell. If it sees abnormal material—fragments of bacteria, cancer proteins, other signs of malfeasance—it can attack. The diversity of the MHC helps mammals here because different MHC proteins fasten on to and raise the alarm against different dangers, so the more diversity in the mammalian MHC, the more things a creature can combat. And crucially, unlike with other traits, MHC genes don't interfere with each other. Mendel identified the first dominant traits, cases where some versions of genes "win out" over others. With the MHC, all the genes work independently, and no one gene masks another. They cooperate; they codominate.

As for its second, more philosophical function, the MHC allows our bodies to distinguish between self and nonself. While mounting protein fragments, MHC genes cause little beard hairs to sprout on the surface of every cell; and because each creature has a unique combination of MHC genes, this cellular beard hair will have a unique arrangement of colors and curls. Any nonself interlopers in the body (like cells from animals or other people) of course have their own MHC genes sprouting their own unique beards. Our immune system is so precise that it will recognize those beards as different, and—even if those cells betray no signs of diseases or parasites—marshal troops to kill the invaders.

Destroying invaders is normally good. But one side effect of the MHC's vigilance is that our bodies reject transplanted organs unless the recipients take drugs to suppress their immune systems. Sometimes even that doesn't work. Transplanting organs from animals could help alleviate the world's chronic shortage of organ donors, but animals have such bizarre (to us) MHCs that our bodies reject them instantly. We even destroy tissues and blood vessels *around* implanted animal organs, like retreating soldiers burning crops so that the enemy can't use them for nourishment either. By absolutely paralyzing the immune system, doctors have kept people alive on baboon hearts and livers for a few weeks, but so far the MHC always wins out.

For similar reasons, the MHC made things difficult for mammal evolution. By all rights, a mammal mother should attack the fetus inside her as a foreign growth, since half its DNA, MHC and otherwise, isn't hers. Thankfully, the placenta mediates this conflict by restricting access to the fetus. Blood pools in the placenta, but no blood actually crosses through to the fetus, just nutrients. As a result, a baby like Emiko should remain perfectly, parasitically invisible to Mayumi's immune cells, and Mayumi's cells should never cross over into Emiko. Even if a few do slip through the placental gate, Emiko's own immune system should recognize the foreign MHC and destroy them.

But when scientists scrutinized the MHC of Mayumi's cancerous blood cells, they discovered something that would be almost admirable in its cleverness, if it weren't so sinister. In humans, the MHC is located on the shorter arm of chromosome six. The scientists noticed that this short arm in Mayumi's cancer cells was even shorter than it should be—because the cells had deleted their MHC. Some unknown mutation had simply wiped it from their genes. This left them functionally invisible on the outside, so neither the placenta nor Emiko's immune cells could classify or recognize them. She had no way to scrutinize them for evidence that they were foreign, much less that they harbored cancer.

Overall, then, scientists could trace the invasion of Mayumi's cancer to two causes: the Philadelphia swap that made them malignant, and the MHC mutation that made them invisible and allowed them to trespass and burrow into Emiko's cheek. The odds of either thing happening were low; the odds of them happening in the same cells, at the same time, in a woman who happened to be pregnant, were astronomically low. But not zero. In fact, scientists now suspect that in most historical cases where mothers gave cancer to their fetuses, something similar disabled or compromised the MHC.

If we follow the thread far enough, the MHC can help illuminate one more aspect of Hideo and Mayumi and Emiko's story, a thread that runs back to our earliest days as mammals. A developing fetus has to conduct a whole orchestra of genes inside every cell, encouraging some DNA to play louder and hushing other sections up. Early on in the pregnancy, the most active genes are the ones that mammals inherited from our egg-laying, lizardlike ancestors. It's a humbling experience to flip through a biology textbook and see how uncannily similar bird, lizard, fish, human, and other embryos appear during their early lives. We humans even have rudimentary gill slits and tails—honest-to-god atavisms from our animal past.

After a few weeks, the fetus mutes the reptilian genes and turns on a coterie of genes unique to mammals, and pretty soon the fetus starts to resemble something you could imagine naming after your grandmother. Even at this stage, though, if the right genes are silenced or tweaked, atavisms (i.e., genetic throwbacks) can appear. Some people are born with the same extra nipples that barnyard sows have.* Most of these extra nipples poke through the "milk line" running vertically down the torso, but they can appear as far away as the sole of the foot. Other atavistic genes leave people with coats of hair sprouting all over their bodies, including their cheeks and foreheads. Scientists can even distinguish (if you'll forgive the pejoratives) between "dog-faced" and "monkey-faced" coats, depending on the coarseness, color, and other qualities of the hair. Infants missing a snippet at the end of chromosome five develop *cri-du-chat*, or "cry of the cat" syndrome, so named for their caterwauling chirps and howls. Some children are also born with tails. These tails— usually centered above their buttocks—contain muscles and nerves and run to five inches long and an inch thick. Sometimes tails appear as side effects of recessive genetic disorders that

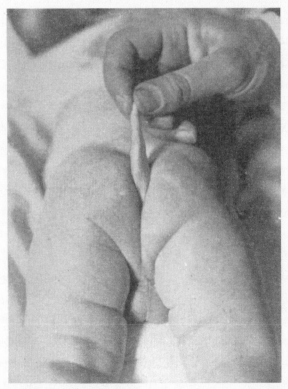

A hale and healthy baby boy born with a tail—a genetic throwback from our primate past. (Jan Bondeson, *A Cabinet of Medical Curiosities*, reproduced by permission)

cause widespread anatomical problems, but tails can appear idio-syncratically as well, in otherwise normal children. Pediatricians have reported that these boys and girls can curl their tails up like an elephant's trunk, and that the tails contract involuntarily when children cough or sneeze.* Again, all fetuses have tails at six weeks old, but they usually retract after eight weeks as tail cells die and the body absorbs the excess tissue. Tails that persist probably arise from spontaneous mutations, but some children with tails do have betailed relatives. Most get the harmless appendage removed just after birth, but some don't bother until adulthood.

All of us have other atavisms dormant within us as well, just waiting for the right genetic signals to awaken them. In fact, there's one genetic atavism that none of us escapes. About forty days after conception, inside the nasal cavity, humans develop a tube about 0.01 inches long, with a slit on either side. This incip-ient structure, the vomeronasal organ, is common among mam-mals, who use it to help map the world around them. It acts like an auxiliary nose, except that instead of smelling things that any sentient creature can sniff out (smoke, rotten food), the vomero-nasal organ detects pheromones. Pheromones are veiled scents vaguely similar to hormones; but whereas hormones give our bod-ies internal instructions, pheromones give instructions (or at least winks and significant glances) to other members of our species.

Because pheromones help guide social interactions, espe-cially intimate encounters, shutting off the VNO in certain mammals can have awkward consequences. In 2007, scientists at Harvard University genetically rewired some female mice to disable their VNOs. When the mice were by themselves, not much changed—they acted normally. But when let loose on regular females, the altered mice treated them like the Romans did the Sabine women. They stormed and mounted the maidens, and despite lacking the right equipment, they began thrusting their hips back and forth. The bizarro females even groaned like men, emitting ultrasonic squeals that, until then, were heard only from male mice at climax.

Humans rely less on scent than other mammals do; throughout our evolution we've lost or turned off six hundred common mammalian genes for smell. But that makes it all the more striking that our genes still build a VNO. Scientists have even detected nerves running from the fetal VNO to the brain, and have seen these nerves send signals back and forth. Yet for unknown reasons, despite going through the trouble of creating the organ and wiring it up, our bodies neglect this sixth sense, and after sixteen weeks it starts shriveling. By adulthood, it has retracted to the point that most scientists dispute whether humans even have a VNO, much less a functional one.

The debate about the human VNO fits into a larger and less-than-venerable historical debate over the supposed links between scent, sexuality, and behavior. One of Sigmund Freud's nuttier friends, Dr. Wilhelm Fliess, classified the nose as the body's most potent sex organ in the late 1800s. His "nasal reflux neurosis theory" was an unscientific hash of numerology, anecdotes about masturbation and menstruation, maps of hypothetical "genital spots" inside the nose, and experiments that involved dabbing cocaine on people's mucus membranes and monitoring their libidos. His failure to actually explain anything about human sexuality didn't lower Fliess's standing; to the contrary, his work influenced Freud, and Freud allowed Fliess to treat his patients (and, some have speculated, Freud himself) for indulging in masturbation. Fliess's ideas eventually died out, but pseudoscientific sexology never has. In recent decades, hucksters have sold perfumes and colognes enriched with pheromones, which supposedly make the scentee a sexual magnet. (Don't hold your breath.) And in 1994 a U.S. military scientist requested $7.5 million from the air force to develop a pheromone-based "gay bomb." His application described the weapon as a "distasteful but completely non-lethal" form of warfare. The pheromones would be sprayed over the (mostly male) enemy troops, and the smell would somehow—the details were tellingly sketchy, at least outside the scientist's fantasies—whip them into such a froth of randiness that they'd drop their weapons and make whoopee instead of war. Our soldiers, wearing gas masks, would simply have to round them up.*

Perfumes and gay bombs aside, some legitimate scientific work has revealed that pheromones can influence human behavior. Forty years ago, scientists determined that pheromones cause the menstrual cycles of women who live together to converge toward the same date. (That's no urban legend.) And while we may resist reducing human love to the interaction of chemicals, evidence shows that raw human lust—or more demurely, attraction—has a strong olfactory component. Old anthropology books, not to mention Charles Darwin himself, used to marvel that in societies that never developed the custom of kissing, potential lovers often sniffed each other instead of smooching. More recently, Swedish doctors ran some experiments that echo the dramatic Harvard study with mice. The doctors exposed straight women, straight men, and homosexual men to a pheromone in male sweat. During this exposure, the brain scans of straight women and gay men—but not straight men—showed signs of mild arousal. The obvious follow-up experiment revealed that pheromones in female urine can arouse straight men and gay women, but not straight women. It seems the brains of people with different sexual orientations respond differently to odors from either sex. This doesn't prove that humans have a functioning VNO, but it does suggest we've retained some of its pheromone-detecting ability, perhaps by genetically shifting its responsibilities to our regular nose.

Probably the most straightforward evidence that smells can influence human arousal comes from—and we've finally circled back to it—the MHC. Like it or not, your body advertises your MHC every time you lift your arm. Humans have a high concentration of sweat glands in the armpit, and mixed in with the excreted water, salt, and oil are pheromones that spell out exactly what MHC genes people have to protect them from disease. These MHC ads drift into your nose, where nasal cells can work out how much the MHC of another person differs from your own. That's helpful in judging a mate because you can estimate the probable health of any children you'd have together. Remember that MHC genes don't interfere with each other—they codominate. So if Mom and Dad have different MHCs, baby will inherit their combined disease resistance. The more genetic disease resistance, the better off baby will be.

This information trickles into our brains unconsciously but can make itself known when we suddenly find a stranger unaccountably sexy. It's impossible to say for sure without testing, but when this happens, the odds are decent that his or her MHC is notably different from your own. In various studies, when women sniffed a T-shirt worn to bed by men they never saw or met, the women rated the men with wild MHCs (compared to their own) as the sexiest in the batch. To be sure, other studies indicate that, in places already high in genetic diversity, like parts of Africa, having a wildly different MHC doesn't increase attraction. But the MHC-attraction link does seem to hold in more genetically homogeneous places, as studies in Utah have shown. This finding might also help explain why—because they have more similar MHCs than average—we find the thought of sex with our siblings repugnant.

Again, there's no sense in reducing human love to chemicals; it's waaaay more complex than that. But we're not as far removed from our fellow mammals as we might imagine. Chemicals do prime and propel love, and some of the most potent chemicals out there are the pheromones that advertise the MHC. If two people from a genetically homogeneous locale—take Hideo and Mayumi—came together, fell in love, and decided to have a child, then as far as we can ever explain these things biologically, their MHCs likely had something to do with it. Which makes it all the more poignant that the disappearance of that same MHC empowered the cancer that almost destroyed Emiko.

Almost. The survival rate for both mothers and infants with simultaneous cancer has remained abysmally low, despite great advances in medicine since 1866. But unlike her mother, Emiko responded to treatment well, partly because her doctors could tailor her chemotherapy to her tumor's DNA. Emiko didn't even need the excruciating bone-marrow transplants that most children with her type of cancer require. And as of today (touch wood) Emiko is alive, almost seven years old and living in Chiba.

We don't think of cancer as a transmissible disease. Twins can nevertheless pass cancer to each other in the womb; transplanted organs can pass cancer to the organ recipient; and mothers can indeed pass cancer to their unborn children, despite the defenses of the placenta. Still, Emiko proves that catching an advanced cancer, even as a fetus, doesn't have to be fatal. And cases like hers have expanded our view of the MHC's role in cancer, and demonstrated that the placenta is more permeable than most scientists imagined. "I'm inclined to think that maybe cells get by [the placenta] in modest numbers all the time," says a geneticist who worked with Emiko's family. "You can learn a lot from very odd cases in medicine."

In fact, other scientists have painstakingly determined that most if not all of us harbor thousands of clandestine cells from our mothers, stowaways from our fetal days that burrowed into our vital organs. Every mother has almost certainly secreted away a few memento cells from each of her children inside her, too. Such discoveries are opening up fascinating new facets of our biology; as one scientist wondered, "What constitutes our psychological self if our brains are not entirely our own?" More personally, these findings show that even after the death of a mother or child, cells from one can live on in the other. It's another facet of the mother-child connection that makes mammals special.

9

Humanzees and Other Near Misses

When Did Humans Break Away from Monkeys, and Why?

od knows the evolution of human beings didn't stop with fur, mammary glands, and placentas. We're also primates—although that was hardly something to brag about sixty million years ago. The first rudimentary primates probably didn't crack one pound or live beyond six years. They probably lived in trees, hopped about instead of striding, hunted nothing bigger than insects, and crept out of their hovels only at night. But these milquetoast midnight bug-biters got lucky and kept evolving. Tens of millions of years later, some clever, opposable-thumbed, chest-beating primates arose in Africa, and members of one of those lines of primates literally rose onto two feet and began marching across the savannas. Scientists have studied this progression intensely, picking it apart for clues about the essence of humanity. And looking back on the whole picture—that *National Geographic* sequence of humans getting up off our knuckles, shedding our body hair, and renouncing our prognathous jaws—we can't help but think about our emergence a little triumphantly.

Still, while the rise of human beings was indeed precious, our DNA—like the slave in Roman times who followed a triumphant general around—whispers in our ears, *Remember, thou art mortal.* In reality the transition from apelike ancestor to modern human being was more fraught than we appreciate. Evidence tattooed into our genes suggests that the human line almost went extinct, multiple times; nature almost wiped us out like so many mastodons and dodos, with nary a care for our big plans. And it's doubly humbling to see how closely our DNA sequence still resembles that of so-called lower primates, a likeness that conflicts with our inborn feeling of preordainment— that we somehow sit superior to other creatures.

One strong piece of evidence for that inborn feeling is the revulsion we feel over the very idea of mixing human tissues with tissues from another creature. But serious scientists throughout history have attempted to make human-animal chimeras, most recently by adulterating our DNA. Probably the all-time five-alarmer in this realm took place in the 1920s, when a Russian biologist named Ilya Ivanovich Ivanov tried to unite human genes with chimpanzee genes in some hair-raising experiments that won the approval of Joseph Stalin himself.

Ivanov started his scientific career around 1900, and worked with physiologist Ivan Pavlov (he of the drooling dogs) before branching out to become the world's expert in barnyard insemination, especially with horses. Ivanov crafted his own instruments for the work, a special sponge to sop up semen and rubber catheters to deliver it deep inside the mares. For a decade, he worked with the Department for State Stud-Farming, an official bureau that supplied the ruling Romanov government with pretty horses. Given those political priorities, it's not hard to imagine why the Romanovs were overthrown in 1917, and when the Bolsheviks took over and founded the Soviet Union, Ivanov found himself unemployed.

It didn't help Ivanov's prospects that most people at the time considered artificial insemination shameful, a corruption of natural copulation. Even those who championed the technique went to ridiculous lengths to preserve an aura of organic sex. One prominent doctor would wait outside a barren couple's room, listening at the keyhole while they went at it, then rush in with a baster of sperm, practically push the husband aside, and spurt it into the woman—all to trick her egg cells into thinking that insemination had happened during the course of inter-course. The Vatican banned artificial insemination for Catho-lics in 1897, and Russia's Greek Orthodox Church similarly condemned anyone, like Ivanov, who practiced it.

But the religious snit eventually helped Ivanov's career. Even while mired in the barnyard, Ivanov had always seen his work in grander terms—not just a way to produce better cows and goats, but a way to probe Darwin's and Mendel's fundamental theories of biology, by mixing embryos from different species. After all, his sponges and catheters removed the main barrier to such work, coaxing random animals to conjugate. Ivanov had been chewing over the ultimate test of Darwinian evolution, human-zees, since 1910, and he finally (after consulting with Hermann Muller, the Soviet-loving *Drosophila* scientist) screwed up the courage to request a research grant in the early 1920s.

Ivanov applied to the people's commissar of enlightenment, the official who controlled Soviet scientific funding. The com-missar, a theater and art expert in his former life, let the proposal languish, but other top Bolsheviks saw something promising in Ivanov's idea: a chance to insult religion, the Soviet Union's avowed enemy. These farsighted men argued that breeding humanzees would be vital "in our propaganda and in our strug-gle for the liberation of the working people from the power of the Church." Ostensibly for this reason, in September 1925— just months after the Scopes trial in the United States—the

Soviet government granted Ivanov $10,000 ($130,000 today) to get started.

Ivanov had good scientific reasons to think the work could succeed. Scientists knew at the time that human and primate blood showed a remarkable degree of similarity. Even more exciting, a Russian-born colleague, Serge Voronoff, was wrapping up a series of sensational and supposedly successful experiments to restore the virility of elderly men by transplanting primate glands and testicles into them. (Rumors spread that Irish poet William Butler Yeats had undergone this procedure. He hadn't, but the fact that people didn't dismiss the rumor as rubbish says a lot about Yeats.) Voronoff's transplants seemed to show that, at least physiologically, little separated lower primates and humans.

Ivanov also knew that quite distinct species can reproduce together. He himself had blended antelopes with cows, guinea pigs with rabbits, and zebras with donkeys. Besides amusing the tsar and his minions (very important), this work proved that animals whose lines had diverged even millions of years ago could still have children, and later experiments by other scientists provided further proof. Pretty much any fantasy you've got—lions with tigers, sheep with goats, dolphins with killer whales—scientists have fulfilled it somewhere. True, some of these hybrids were and are sterile, genetic dead ends. But only some: biologists find many bizarre couplings in the wild, and of the more than three hundred mammalian species that "outbreed" naturally, fully one-third produce fertile children. Ivanov fervently believed in crossbreeding, and after he sprinkled some good old Marxist materialism into his calculations—which denied human beings anything as gauche as a soul that might not condescend to commingle with chimps—then his humanzee experiments seemed, well, doable.

Scientists don't know even today whether humanzees, however icky and unlikely, are at least possible. Human sperm can

A modern zonkey—a zebra-donkey mix. Ilya Ivanov created zonkeys (which he called "zeedonks") and many other genetic hybrids before pursuing human-zees. (Tracy N. Brandon)

pierce the outer layer of some primate eggs in the lab, the first step in fertilization, and human and chimpanzee chromosomes look much the same on a macro scale. Heck, human DNA and chimp DNA even enjoy each other's company. If you prepare a solution with both DNAs and heat it up until the double strands unwind, human DNA has no problem embracing chimp DNA and zipping back up with it when things cool down. They're that similar.*

What's more, a few primate geneticists think that our ancestors resorted to breeding with chimps long *after* we'd split away to become a separate species. And according to their controversial but persistent theory, we copulated with chimps far longer than most of us are comfortable thinking about, for a million years. If true, our eventual divergence from the chimp line was a complicated and messy breakup, but not inevitable. Had things gone another way, our sexual proclivities might well have rubbed the human line right out of existence.

The theory goes like this. Seven million years ago some unknown event (maybe an earthquake opened a rift; maybe half the group got lost looking for food one afternoon; maybe a bitter butter battle broke out) split a small population of primates. And with every generation they remained apart, these two separate groups of chimp-human ancestors would have accumulated mutations that gave them unique characteristics. So far, this is standard biology. More unusually, though, imagine that the two groups reunited some time later. Again, the reason is impossible to guess; maybe an ice age wiped out most of their habitats and squeezed them together into small woodland refugia. Regardless, we don't need to propose any outlandish, Marquis de Sade motivations for what happened next. If lonely or low in numbers, the protohumans might eagerly—despite having forsworn the comforts of protochimps for a million years—have welcomed them back into their beds (so to speak) when the groups reunited. A million years may seem like forever, but the two protos would have been less distinct genetically than many interbreeding species today. So while this interbreeding might have produced some primate "mules," it might have produced fertile hybrids as well.

Therein lay the danger for protohumans. Scientists know of at least one case in primate history, with macaques, when two long-separate species began mating again and melded back into one, eliminating any special differences between them. Our interbreeding with chimps was no weekend fling or dalliance; it was long and involved. And if our ancestors had said what the hell and settled down with protochimpanzees permanently, our unique genes could have drowned in the general gene pool in the same way. Not to sound all eugenicky, but we would have humped ourselves right out of existence.

Of course, this all assumes that chimps and humans did revert to sleeping together after an initial split. So what's the evidence for this charge? Most of it lies on our (wait for it) sex chromosomes, especially the X. But it's a subtle case.

When female hybrids have fertility trouble, the flaw usually traces back to their having one X from one species, one X from another. For whatever reason, reproduction just doesn't go as smoothly with a mismatch. Mismatched sex chromosomes hit males even harder: an X and a Y from different species almost always leave them shooting blanks. But infertility among women is a bigger threat to group survival. A few fertile males can still impregnate loads of females, but no gang of fertile males can make up for low female fecundity, because females can have children only so quickly.

Nature's solution here is genocide. That is, gene-o-cide: nature will eliminate any potential mismatches among the inter-breeders by eradicating the X chromosome of one species. It doesn't matter which, but one has to go. It's a war of attrition, really. Depending on the messy details of how many proto-chimps and protohumans interbred, and then whom exactly the first generation of hybrids reproduced with, and then their dif-ferential birthrates and mortality—depending on all that, one species' X chromosomes probably appeared in higher numbers initially in the gene pool. And in the subsequent generations, the X with the numbers advantage would slowly strangle the other one, because anyone with similar Xs would outbreed the half-breeds.

Notice there's no comparable pressure to eliminate nonsex chromosomes. Those chromosomes don't mind being paired with chromosomes from the other species. (Or if they do mind, their quarrel likely won't interfere with making babies, which is what counts to DNA.) As a result, the hybrids and their descendants could have been full of mismatched nonsex chromosomes and survived just fine.

Scientists realized in 2006 that this difference between sex and nonsex chromosomes might explain a funny characteristic of human DNA. After the initial split between their lines, protochimps and protohumans should have started down different paths and accumulated different mutations on each chromosome. And they did, mostly. But when scientists look at chimps and humans today, their Xs look more uniform than other chromosomes. The DNA clock on X got reset, it seems; it retained its girlish looks.

We hear the statistic sometimes that we have 99 percent of our DNA coding region in common with chimps, but that's an average, overall measure. It obscures the fact that human and chimp Xs, a crucial chromosome for Ivanov's work, look even more identical up and down the line. One parsimonious way to explain this similarity is interbreeding and the war of attrition that would probably have eliminated one type of X. In fact, that's why scientists developed the theory about protohuman and protochimp mating in the first place. Even they admit it sounds a little batty, but they couldn't puzzle out another way to explain why human and chimp X chromosomes have less variety than other chromosomes.

Fittingly, however (given the battle of the sexes), research related to Y chromosomes may contradict the X-rated evidence for human-chimp interbreeding. Again, scientists once believed that Y—which has undergone massive shrinkage over the past 300 million years, down to a chromosomal stub today—would one day disappear as it continued to shed genes. It was considered an evolutionary vestige. But in truth Y has evolved rapidly even in the few million years since humans swore off chimpanzees (or vice versa). Y houses the genes to make sperm, and sperm production is an area of fierce competition in wanton species. Many different protogents would have had sex with each protolady, so one gent's sperm constantly had to wrestle with another gent's inside her vagina. (Not appetizing, but true.) One evolutionary strategy to secure an advantage here is to produce loads and loads of sperm each time you ejaculate. Doing so of course requires copying and pasting lots of DNA, because each sperm needs its own genetic payload. And the more copying that takes place, the more mutations that occur. It's a numbers game.

However, these inevitable copying mistakes plague the X chromosome less than any other chromosome because of our reproductive biology. Just like making sperm, making an egg requires copying and pasting lots of DNA. A female has equal numbers of every chromosome: two chromosome ones, two chromosome twos, and so on, as well as two Xs. So during the production of eggs, each chromosome, including the Xs, gets copied equally often. Males also have two copies of chromosomes one through twenty-two. But instead of two Xs, they have one X and one Y. During the production of sperm, then, the X gets copied less often compared to other chromosomes. And because it gets copied less often, it picks up fewer mutations. That mutation gap between X and other chromosomes widens even more when—because of Y chromosome–fueled sperm competition—males begin churning out loads of sperm. Therefore, some biologists argue, the seeming lack of mutations on X when comparing chimps and humans might not involve an elaborate and illicit sexual history. It might result from our basic biology, since X should always have fewer mutations.*

Regardless of who's right, work along these lines has undermined the old view of the Y as a misfit of the mammal genome; it's quite sophisticated in its narrow way. But for humans, it's hard to say if the revisionary history is for the better. The pressure to develop virile sperm is much higher in chimps than humans because male chimps have more sex with different partners. In response, evolution has remade the chimp Y thoroughly top to bottom. So thoroughly, in fact, that—contrary to what most men probably want to believe—chimps have pretty much left us guys in the dust evolutionarily. Chimps simply have tougher, smarter swimmers with a better sense of direction, and the human Y looks obsolete in comparison.

But that's DNA for you—humbling. As one Y-chromosome specialist comments, "When we sequenced the chimp genome, people thought we'd understand why we have language and write poetry. But one of the most dramatic differences turns out to be sperm production."

Talking about Ivanov's experiments in terms of DNA is a little anachronistic. But scientists in his day did know that chromosomes shuttled genetic information down through the generations and that chromosomes from mothers and fathers had to be compatible, especially in numbers. And based on the preponderance of evidence, Ivanov decided that chimps and humans had similar enough biologies to push forward.

After securing funds, Ivanov arranged through a colleague in Paris to work at a primate research station in colonial French Guinea (modern Guinea). The conditions at the station were deplorable: the chimps lived in cages exposed to the weather, and half of the seven hundred chimps that local poachers had netted and dragged back had died of disease or neglect. Nevertheless Ivanov enlisted his son (another triple *i:* Ilya Ilich Ivanov) and began chugging thousands of nautical miles back and forth between Russia, Africa, and Paris. The Ivanovs eventually arrived in sultry Guinea, ready to begin experiments, in November 1926.

Because the captives were sexually immature, too young to conceive, Ivanov wasted months furtively checking their pubic fur each day for menstrual blood. Meanwhile fresh inmates kept piling in, right up until Valentine's Day 1927. Ivanov had to keep his work secret to avoid angry questions from the Guineans, who had strong taboos against humans mating with primates, based on local myths about hybrid monsters. But finally, on February 28, two chimps named Babette and Syvette had their periods. At eight o'clock the next morning, after visiting an anonymous local donor, Ivanov and son approached their cages armed with a syringe of sperm. They were also armed with two Browning pistols: Ivanov Jr. had been bitten and hospitalized a few days before. The Ivanovs didn't end up needing the Brownings, but only because they more or less raped the chimps, binding Babette and Syvette in nets. The virgin chimps thrashed about anyway, and Ivanov managed to jam the syringe only into their vaginas, not into their uteruses, the preferred spot for deposits. Not surprisingly, this experiment failed, and Babette and Syvette got their periods again weeks later. Many juvenile chimps died of dysentery in the following months, and Ivanov managed to inseminate just one more female that spring (she was drugged). This attempt also failed, meaning Ivanov had no living human-zees to haul back to the Soviet Union to win more funding.

Perhaps wary that the commissar of enlightenment might not give him a second chance, Ivanov began pursuing new projects and new avenues of research, some in secret. Before leaving for Africa, he'd helped break ground on the first Soviet primate station, in Sukhumi, in modern-day Georgia, one of the few semitropical areas in the Soviet empire, and conveniently located in the homeland of the new Soviet leader, Joseph Stalin. Ivanov also courted a wealthy but flaky Cuban socialite named Rosalià Abreu, who operated a private primate sanctuary on her estate in Havana—partly because she believed that chimpanzees had psychic powers and deserved protection. Abreu initially agreed to house Ivanov's experiments but withdrew her offer because she feared newspapers would get hold of the story. She was right to worry. The *New York Times* caught wind of the story anyway after some of Ivanov's American supporters appealed to the publicity-hungry American Association for the Advancement of Atheism for funding and it began trumpeting the idea. The *Times* story* prompted the Ku Klux Klan to send out letters warning Ivanov about doing his devil's work on their side of the Atlantic, as it was "abominable to the Creator."

Meanwhile Ivanov was finding it expensive and annoying to keep a large chimp harem healthy and safe, so he devised a plan to turn his experimental protocol inside out. Females in primate societies can have children only so quickly, while one virile male can spread his seed widely at little cost. So instead of keeping multiple female chimps around to impregnate with human sperm, Ivanov decided to secure one male chimp and impregnate human women. Simple as that.

To this end, Ivanov made secret contact with a colonial doctor in the Congo and asked the doctor to let him inseminate patients. When the doctor asked why on earth his patients would consent, Ivanov explained that they wouldn't tell them. This satisfied the doctor, and Ivanov beat cheeks from French Guinea down to the Congo, where everything seemed ready to go. At the last second, however, the local governor intervened and informed Ivanov that he could not perform the experiment in the hospital: he had to do it outdoors. Offended at the interference, Ivanov refused; the unsanitary conditions would compromise his work and the safety of his patients, he said. But the governor held firm. Ivanov called the debacle a "terrible blow" in his diary.

Until his last day in Africa, Ivanov kept hunting down leads for other women to inseminate, but nothing came of them. So when he finally left French Guinea in July 1927, he decided no more fooling around in far-flung places. He would shift his operations to the newly opened Soviet primate station at Sukhumi. He would also sidestep the hassles of female chimp harems by finding one reliable male stud and coaxing Soviet women to reproduce with him.

As feared, Ivanov had trouble finding funding for his revised experiments, but the Society of Materialist Biologists took up Ivanov's work as a proper Bolshevik cause and ponied up. Before he could get started, however, most of the Sukhumi primates fell sick and died that winter. (Though balmy by Soviet standards, Sukhumi was pretty far north for African primates.) Luckily, the one primate to pull through was male—Tarzan, a twenty-six-year-old orangutan. All Ivanov needed now were human recruits. Officials, though, informed Ivanov he could not offer the recruits money for being surrogates. They had to volunteer, out of ideological amour for the Soviet state, a less tangible reward than a bounty. This further delayed things, but nevertheless, by spring 1928, Ivanov had his mark.

Her name comes down to us only as "G." Whether she was slender or stout, freckled or fair, a madame or a charwoman, we have no idea. All we have is a heartrending and elliptical letter she wrote to Ivanov: "Dear Professor[:] With my private life in ruins, I don't see any sense in my further existence.... But when I think that I could do a service for science, I feel enough courage to contact you. I beg you, don't refuse me."

Ivanov assured her he wouldn't. But as he made arrangements to bring G to Sukhumi and inseminate her, Tarzan died of a brain hemorrhage. No one had had time to milk him for sperm, either. Once again the experiment stalled.

This time, permanently. Before he could round up another ape, the Soviet secret police arrested Ivanov in 1930 for obscure reasons and exiled him to Kazakhstan. (The official charge was that old standby, "counterrevolutionary activities.") Already past sixty, Ivanov took to prison about as well as his primates had, and

Soviet biologist Ilya Ivanovich Ivanov went further than any scientist ever has in trying to breed primates with humans. (Institute of the History of Natural Sciences and Technology, Russian Academy of Sciences)

his health grew fragile. He was cleared of the bogus charges in 1932, but the day before he was to leave prison, he, like Tarzan, had a brain hemorrhage. Within a few days, he joined Tarzan in that giant primate research station in the sky.

After Ivanov died, his scientific agenda disintegrated. Few other scientists had the skill to inseminate primates, and just as important, no scientifically advanced country was as willing as the Soviet Union to trash every ethical guideline out there and fund such work. (Although to be fair, even hardened Politburo officials threw up in their mouths when Ivanov revealed his clandestine attempt to impregnate hospitalized women in the Congo with chimp sperm.) As a result, since the 1920s scientists have done virtually no research on human-primate hybrids. Which means that Ivanov's most pressing question remains open: could G and a beast like Tarzan have produced a child?

In one way, maybe. In 1997 a biologist in New York applied for a patent on a process to mix embryonic cells from humans and chimps and gestate them in a surrogate mother. The biologist considered the project technically feasible, even if he himself never intended to make a humanzee chimera; he merely wanted to prevent some nefarious person from obtaining a patent first. (The patent office turned the claim down in 2005 because, among other things, patenting a half human could violate the Thirteenth Amendment's prohibition against slavery and owning another human.) However, the process would have required no actual hybridization—no actual mixing of the two species' DNA. That's because the chimp and human embryonic cells would have come into contact only after fertilization; so each individual cell in the body would retain its wholly chimp or wholly human nature. The creature would have been a mosaic, not a hybrid.

Nowadays scientists could easily splice bits of human DNA into chimp embryos (or vice versa), but this would be little more than a tweak biologically. True hybridization requires the old-fashioned fifty-fifty mingling of sperm and eggs, and almost all respectable scientists today would lay money on human-chimp fertilization being impossible. For one, the molecules that form a zygote and start it dividing are specific to each species. And even if a viable humanzee zygote did form, humans and chimps regulate DNA very differently. So the job of getting all that DNA to cooperate and turn genes off and on in sync and make proper skin and liver and especially brain cells would be daunting.

Another reason to doubt that humans and chimps could produce children is the different chromosome count in the two species, a fact that emerged only after Ivanov's time. Getting an accurate chromosome count was surprisingly tricky for most of the twentieth century. DNA remains quite tangled inside the nucleus except for the few moments right before a cell divides, when compact chromosomes form. Chromosomes also have a bad habit of melting together after cells die, which makes counting harder still. Counting is therefore easiest in recently living samples of cells that divide often—like the sperm-making cells inside male gonads. Finding fresh monkey testes wasn't too onerous even in the early 1900s (Lord knows they killed enough monkeys then), and biologists determined that close primate relatives like chimps, orangutans, and gorillas all had forty-eight chromosomes. But lingering taboos made obtaining human testes more difficult. People didn't donate their bodies to science back then, and some desperate biologists—much like those anatomists in the Renaissance who robbed graves—took to lurking near the town gallows to harvest the testicles of condemned criminals. There was simply no other way to get fresh samples.

Given the difficult circumstances, work on human chromosome number remained sketchy; guesses ranged from sixteen to fifty-some. And despite the constant counts in other species, some European scientists beholden to racial theories proclaimed that Asian, black, and white people clearly had different numbers of chromosomes. (No points for guessing who they thought had most.) A Texas biologist named Theophilus Painter—who later discovered the gigantic salivary chromosomes in fruit flies—finally killed the theory of varying chromosome number with a definitive study in 1923. (Rather than depend on the criminal-justice system for raw material, Painter counted his blessings that a former student worked at a lunatic asylum and had access to freshly castrated inmates.) But even Painter's best slides showed human cells with either forty-six or forty-eight chromosomes, and after going round and round and counting and recounting them from every which angle, Painter still couldn't decide. Perhaps worried his paper would be rejected if he didn't at least pretend to know, Painter acknowledged the confusion, took a breath, and guessed—wrongly. He said that humans have forty-eight, and that became the standard figure.

After three decades and the invention of far better microscopes (not to mention an easing of restrictions on human tissues), scientists had rectified the boner, and by 1955 they knew that humans had forty-six chromosomes. But as so often happens, deposing one mystery just inaugurated another, because now scientists had to figure out how humans ended up two chromosomes down anyway.

Surprisingly, they determined that the process started with something like a Philadelphia swap. Around a million years ago, in some fateful man or woman, what were the twelfth and thirteenth human chromosomes (and still are the twelfth and thirteenth chromosomes in many primates) entwined their arms at the very tip, to try and swap material. But instead of separating cleanly, twelve and thirteen got stuck. They fused together at their ends, like one belt buckled onto another. This amalgam eventually became human chromosome number two.

Fusions like this are actually not uncommon—they occur once every thousand births—and most tip-to-tip fusions go unnoticed because they don't upset anyone's health. (The ends of chromosomes often contain no genes, so nothing gets disrupted.) However, notice that a fusion by itself can't explain the drop from forty-eight to forty-six. A fusion leaves a person with forty-seven chromosomes, not forty-six, and the odds of two identical fusions in the same cell are pretty remote. And even after the drop to forty-seven, the person still has to pass his genes on, a serious barrier.

Scientists did eventually puzzle out what must have happened. Let's go back a million years, when most protohumans had forty-eight chromosomes, and follow a hypothetical Guy, who has forty-seven. Again, a chromosome fused at the tips won't affect Guy's day-to-day health. But having an odd number of chromosomes will cripple the viability of his sperm, for a simple reason. (If you prefer to think of a female, the same arguments apply to her eggs.) Say the fusion left Guy with one normal chromosome 12, one normal 13, and a 12-13 hybrid. During sperm production, his body has to divide those three chromosomes into two cells at one point, and if you do the math, there are only a few possible ways to divvy them up. There's {12} and {13, 12-13}, or {13} and {12, 12-13}, or {12, 13} and {12-13}. The first four sperm either are missing a chromosome or have a duplicate, practically a cyanide capsule for an embryo. The last two cases have the proper amount of DNA for a normal child. But only in the sixth case does Guy pass the fusion on. Overall, then, because of the odd number, two-thirds of Guy's children die in the womb, and just one-sixth inherit the fusion. But any Junior with the fusion would then face the same terrible odds trying to reproduce. Not a good recipe for spreading the fusion—and again, that's still only forty-seven chromosomes, not forty-six.

What Guy needs is a Doll with the same two fused chromosomes. Now, the odds of two people with the same fusion meeting and having children might seem infinitesimal. And they would be—except in inbred families. Relatives share enough genes that, given one person with a fusion, the chances of finding a cousin or half sibling with the same fusion don't round down to zero so easily. What's more, while the odds of Guy and Doll having a healthy child remain low, every thirty-sixth spin of the genetic roulette wheel (because $1/6 \times 1/6 = 1/36$), the child would inherit *both* fused chromosomes, giving him forty-six total. And here's the payoff: Junior and his forty-six chromosomes would have a much easier time having children. Remember that the fusion itself doesn't disable or ruin a chromosome's

DNA; lots of healthy people worldwide have fusions. It's only reproduction that gets tricky, since fusions can lead to an excess or deficit of DNA in embryos. But because he has an even number of chromosomes, little Junior wouldn't have any unbalanced sperm cells: each would have exactly the right amount of DNA to run a human, just packaged differently. As a result, all his children would be healthy. And if his children start having their own children—especially with other relatives who have forty-six or forty-seven chromosomes—the fusion could start to spread.

Scientists know this scenario isn't just hypothetical, either. In 2010 a doctor in rural China discovered a family with a history of consanguineous (similar-blood) marriages. And among the various overlapping branches of the family tree, he discovered a male who had forty-four chromosomes. In this family's case, the fourteenth and fifteenth chromosomes had fused, and consistent with the example of Guy and Doll, they had a brutal record of miscarriages and spontaneous abortions in their past. But from that wreckage, a perfectly healthy man with two fewer chromosomes emerged—the first known stable reduction since our ancestors started down the path to forty-six chromosomes a million years ago.*

In one sense, then, Theophilus Painter was right: for most of our primate history, the human line did have the same number of chromosomes as many primates. And until that transition, the hybrids that Ivanov coveted were far more possible. Having a different number of chromosomes won't always prevent breeding; horses have sixty-four chromosomes, and donkeys sixty-two. But again, molecular gears and cogs don't turn nearly as smoothly when chromosomes don't correspond. Indeed, it's telling that Painter published his study in 1923, just before Ivanov started his experiments. Had Painter guessed forty-six instead of forty-eight, that might have been a serious blow to Ivanov's hopes.

And perhaps not just Ivanov's. The issue remains disputed,

and most historians dismiss the story as legend, if not outright legerdemain. But according to documents that a Russian historian of science unearthed in Soviet archives, Joseph Stalin himself approved the funding for Ivanov's work. This is strange, since Stalin abhorred genetics: he later allowed his scientific hatchet man Trofim Lysenko to outlaw Mendelian genetics in the Soviet Union, and, poisoned by Lysenko's influence, angrily rejected Hermann Muller's eugenics program to breed better Soviet citizens. (Muller fled in response, and colleagues he left behind were shot as "enemies of the people.") And that discrepancy—supporting Ivanov's indecent proposals, yet rejecting Muller's so vehemently—has led a few Russian historians to suggest (and here's the dubious bit) that Stalin dreamed of using Ivanov's humanzees as slaves. The legend really took off in 2005 when, through a series of convoluted attributions, the *Scotsman* newspaper in Great Britain quoted unnamed Moscow newspapers quoting still more recovered documents that themselves supposedly quoted Stalin as saying: "I want a new invincible human being, insensitive to pain, resistant and indifferent about the quality of food they eat." On the same day, the *Sun* tabloid also quoted Stalin as saying he thought it best if humanzees have "immense strength but...an underdeveloped brain," presumably so they wouldn't revolt or be miserable enough to kill themselves. Apparently Stalin coveted the beasts to build his Trans-Siberian Railway through Gulag-land, one of history's all-time biggest boondoggles, but his primary goal was repopulating the Red Army, which in World War I (as in most Russian wars) had suffered massive losses.

It's a fact that Stalin approved funding for Ivanov. But not much, and he approved funding for hundreds of other scientists as well. And I've seen no firm evidence—or any evidence, really—that Stalin lusted after a humanzee army. (Nor that he planned, as some suggest, to seek immortality by harvesting humanzee glands

and transplanting them into himself and other top Kremlin officials.) Still, I have to admit it's a hell of a lot of fun to speculate
about this. If Stalin did take a creepy interest in Ivanov's work,
that might explain why Ivanov got funding just as Stalin consoli-
dated his power and decided to rebuild the military. Or why Ivanov established the primate station in Georgia, Stalin's
homeland. Or why the secret police arrested Ivanov after his
failures, and why Ivanov couldn't pay surrogate mothers but had
to find volunteers who would reproduce for the love of Mother
Russia—because after nursing, they would turn their "sons"
and "daughters" over to Papa Stalin anyway. The international
counterfactuals are even more fascinating. Would Stalin have
sent monkey battalions over the North Pole, to invade North
America? Would Hitler still have signed the nonaggression pact
if he'd known Stalin was polluting the Caucasian race like this?

Still, assuming Ivanov could even create humanzees, Papa's
purported military plans would probably have come to naught.
If nothing else—and I'm ignoring the difficulty of training half
chimps to drive tanks or shoot Kalashnikovs—the Soviet Union's
climate alone would probably have annihilated them. Ivanov's
primates suffered from being too far north on the palm-treed
coast of Georgia, so it seems doubtful that even a hybrid would
have survived Siberia or months of trench warfare.*

What Stalin really needed weren't humanzees but
Neanderthals—big, bad, hairy hominids adapted to icy weather.
But of course Neanderthals had gone extinct tens of thousands
of years before, for reasons that remain blurry. Some scientists
once believed that we actively drove Neanderthals into extinction through war or genocide. That theory has fallen out of
favor, and theories about competition over food or climate
change have come to the fore. But in all likelihood, there was
nothing inevitable about our survival and their death. In fact,
for much of our evolution, we humans were probably as dainty

and vulnerable as Ivanov's primates: cold snaps, habitat loss, and natural disasters seem to have crashed our population numbers time and again. And far from this being distant history, we're still dealing with the repercussions. Notice that we've once again explained one mystery of human DNA—how an inbred family might drop two chromosomes—only to raise another—how that new DNA became standard in all humans. It's possible that the ancient twelve-thirteen fusion created fancy new genes, giving the family survival advantages. But probably not. A more plausible explanation is that we suffered a genetic bottleneck—that something wiped out everyone on earth except a few tribes, and that whatever genes those dumb-lucky survivors had spread far and wide. Some species get caught in bottlenecks and never escape—behold Neanderthals. As the scars in our DNA attest, we human beings scraped through some pretty narrow bottlenecks ourselves, and might easily have joined our thick-browed brethren in Darwin's dustbin.

PART III

Genes and Geniuses

How Humans Became All Too Human

10

Scarlet A's, C's, G's, and T's

Why Did Humans Almost Go Extinct?

C risp mice in golden batter. Panther chops. Rhino pie. Trunk of elephant. Crocodile for breakfast. Sliced porpoise head. Horse's tongue. Kangaroo ham.

Yes, domestic life was a trifle off at William Buckland's. Some of his Oxford houseguests best remembered the front hallway, lined like a catacomb with the grinning skulls of fossilized monsters. Others remembered the live monkeys swinging around, or the pet bear dressed in a mortarboard cap and academic robes, or the guinea pig nibbling on people's toes beneath the dinner table (at least until the family hyena crushed it one afternoon). Fellow naturalists from the 1800s remembered Buckland's bawdy lectures on reptile sex (though not always fondly; the young Charles Darwin thought him a buffoon, and the *London Times* sniffed that Buckland needed to watch himself "in the presence of ladies"). And no Oxonian ever forgot the performance art stunt he pulled one spring when he wrote "G-U-A-N-O" on the lawn with bat feces, to advertise it as fertilizer. The word did indeed blaze green all summer.

But most people remembered William Buckland for his diet.

William Buckland ate his way through most of the animal kingdom. (Antoine Claudet)

A biblical geologist, Buckland held the story of Noah's ark dear, and he ate his way through most of Noah's litter, a habit he called "zoophagy." Any flesh or fluid from any beast was eligible for ingestion, be it blood, skin, gristle, or worse. While touring a church once, Buckland startled a local vicar—who was showing off the miraculous "martyr's blood" that dripped from the rafters every night—by dropping to the stone floor and dabbing the stain with his tongue. Between laps Buckland announced, "It's bat urine." Overall Buckland found few animals he couldn't stomach: "The taste of mole was the most repulsive I knew," he once mused. "Until I tasted a bluebottle [fly]."*

Buckland may have hit upon zoophagy while collecting fossils in some remote pocket of Europe with limited dining options.

It may have been a harebrained scheme to get inside the minds of the extinct animals whose bones he dug up. Mostly, though, he just liked barbecuing, and he kept up his hyper-carnivorous activities well into old age. But in one sense, the most amazing thing about Buckland's diet wasn't the variety. It was that Buckland's intestines, arteries, and heart could digest so much flesh, period, and not harden over the decades into a nineteenth-century Body Worlds exhibit. Our primate cousins could never survive the same diet, not even close.

Monkeys and apes have molars and stomachs adapted to pulping plant matter, and eat mostly vegan diets in the wild. A few primates, like chimpanzees, do eat a few ounces of termites or other animals each day on average, and boy do they love tucking into small, defenseless mammals now and then. But for most monkeys and apes, a high-fat, high-cholesterol diet trashes their insides, and they deteriorate at sickening speeds compared to modern humans. Captive primates with regular access to meat (and dairy) often end up wheezing around inside their cages, their cholesterol pushing 300 and their arteries paved with lard. Our protohuman ancestors certainly also ate meat: they left too many stone cleavers lying next to piles of megamammal bones for it all to be coincidence. But for eons early humans probably suffered no less than monkeys for their love of flesh—Paleolithic Elvises wandering the savanna.

So what changed between then and now, between Grunk in ancient Africa and William Buckland at Oxford? Our DNA. Twice since we split off from chimps, the human *apoE* gene has mutated, giving us distinct versions. Overall it's the strongest candidate around (though not the only candidate) for a human "meat-eating gene." The first mutation boosted the performance of killer blood cells that attack microbes, like the deadly microbes lingering in mouthfuls of raw flesh. It also protected against chronic inflammation, the collateral tissue damage that occurs when microbial infections never quite clear up. Unfortunately this *apoE* probably mortgaged our long-term health for short-term gain: we could eat more meat, but it left our arteries looking like the insides of Crisco cans. Lucky for us, a second mutation appeared 220,000 years ago, which helped break nasty fats and cholesterol down and spared us from premature decrepitude. What's more, by sweeping dietary toxins from the body, it kept cells fitter and made bones denser and tougher to break in middle age, further insurance against early death. So even though early humans ate a veritable Roman-orgy diet compared to their fruitarian cousins, *apoE* and other genes helped them live twice as long.

Before we congratulate ourselves, though, about getting our hands on better *apoE*s than monkeys, a few points. For starters, bones with hack marks and other archaeological evidence indicate that we started dining on meat eons before the cholesterol-fighting *apoE* appeared, at least 2.5 million years ago. So for millions of years we were either too dim to link eating meat and early retirement, too pathetic to get enough calories without meat, or too brutishly indulgent to stop sucking down food we knew would kill us. Even less flattering is what the germicidal properties of the earlier *apoE* mutation imply. Archaeologists have found sharpened wooden spears from 400,000 years ago, so some caveman studs were bringing home bacon by then. But what about before that? The lack of proper weapons, and the fact that *apoE* combats microbes—which thrive in shall we say less-than-fresh cuts of carrion—hint that protohumans scavenged carcasses and ate putrid leftovers. At best, we waited for other animals to fell game, then scared them off and stole it, hardly a gallant enterprise. (At least we're in good company here. Scientists have been having the same debate for some time about *Tyrannosaurus rex:* Cretaceous alpha-killer, or loathsome poacher?)

Once again DNA humbles and muddies our view of ourselves. And *apoE* is just one of many cases where DNA research has transformed our knowledge of our ancient selves: filling in forgotten details in some narratives, overthrowing long-held beliefs in others, but always, always revealing how fraught hominid history has been.

To appreciate how much DNA can supplement, annotate, or plain rewrite ancient history, it helps to look back to the days when scholars first started digging up human remains and studying them—the beginnings of archaeology and paleontology. These scientists started off with confidence about human origins, were thrown into confusion by unsettling finds, and only recently flowed back toward (if not all the way to) clarity, thanks largely to genetics.

Except in unnatural cases, like Dutch sailors massacring dodos, virtually no scientist before 1800 believed that species went extinct. They had been created as is, and that was that. But a French naturalist named Jean-Léopold-Nicolas-Frédéric Cuvier upended this notion in 1796. Cuvier was a formidable man, half Darwin, half Machiavelli. He later latched onto Napoleon and rode the little dictator's blue coattails to the pinnacle of European scientific power; by life's end, he was Baron Cuvier. But along the way the baron proved himself one of the great naturalists ever (his power wasn't undeserved), and he built an authoritative case that species could in fact vanish. The first clue came when he recognized that an ancient pachyderm, unearthed in a quarry near Paris, had no living descendants. Even more spectacularly, Cuvier disproved old legends about the so-called *Homo diluvii testis* skeleton. These bones, unearthed years before in Europe, resembled a deformed man with stunted limbs. Folklore had identified "him" as one of the lecherous and corrupted folk that God had expunged with Noah's flood. The less credulous

Cuvier correctly identified the skeleton as, of all things, a titanic salamander that had long ago disappeared from the earth.

Still, not everyone believed Cuvier about the impermanence of species. The keen amateur naturalist (and U.S. president) Thomas Jefferson instructed Lewis and Clark to keep their eyes peeled inside the Louisiana Territory for giant sloths and mastodons. Fossils of both creatures had previously turned up in North America, drawing huge crowds to dig sites. (Charles Willson Peale's painting *The Exhumation of the Mastodon* captures the scene elegantly.) Jefferson wanted to track down living examples of these beasts for patriotic reasons: he was fed up with

The Exhumation of the Mastodon, by Charles Willson Peale, showing the discovery of mastodon bones in New York in 1801. U.S. president Thomas Jefferson argued that mastodons must still be lumbering across North America, and ordered Lewis and Clark to keep their eyes peeled. (MA5911, courtesy of the Maryland Historical Society)

European naturalists who, without ever coming within an ocean of America, dismissed the fauna here as sickly, weak, and stunted, a snobby theory called "American degeneracy." Jefferson wanted to prove American wildlife was as big and hairy and virile as European beasts, and underlying his hope that mastodons and giant sloths still roamed (or inched across) the Great Plains was the belief that species cannot go extinct.

Although William Buckland came down more with the sober extinctionists than the excitable nonextinctionists, he contributed to the debate in his characteristically flamboyant way. For his honeymoon, Buckland dragged his wife specimen hunting across Europe; and even while hiking to remote outcroppings and pickaxing rocks for fossils, he insisted on wearing black academic robes and often a top hat. In addition to bones, Buckland grew obsessed with fossilized hunks of animal shit, called coprolites, which he generously donated to museums. But Buckland made discoveries exciting enough to be forgiven his eccentricities. In one case he excavated an ancient underground predators' lair in Yorkshire, with plenty of snarly teeth and gnawed-on skulls to wow the public. But the work had great scientific merit and bolstered the extinctionist case: the predators were cave hyenas, and since those hyenas no longer lived in England, they must have gone extinct. More profoundly—and fitting, given his proclivity for meat—Buckland identified some vast bones exhumed from an English quarry as a new species of giant reptile, the first example of the most terrifying carnivores ever, dinosaurs. He named it *Megalosaurus.**

However confident he was with extinct animals, Buckland wavered, even equivocated, on the more loaded question of whether ancient human lineages ever existed. Although an ordained minister, Buckland didn't believe in the aleph-by-aleph accuracy of the Old Testament. He speculated that geological eras had existed before "In the Beginning," eras populated with the likes of *Megalosaurus*. Nevertheless, like virtually all scientists, Buckland hesitated to contradict Genesis regarding human origins and our special, recent creation. In 1823, when Buckland unearthed the alluring Red Lady of Paviland—a skeleton covered with seashell jewelry and dusted with red ocher makeup—he ignored plenty of contextual evidence and identified her as a witch or prostitute from no earlier than Roman times. The lady was actually thirty thousand years old (and a man). Buckland also dismissed clear evidence at another site of chipped-flint tools appearing in the same soil layer as pre-Genesis beasts like mammoths and saber-toothed tigers.

Even less forgivable, Buckland pretty much dropped a steaming coprolite on one of the most spectacular archaeological discoveries ever. In 1829 Philippe-Charles Schmerling unearthed, among some ancient animal remains, a few uncanny, human-but-not-quite-human bones in Belgium. Basing his conclusions especially on skull fragments from a child, he suggested they belonged to an extinct hominid species. Buckland examined the bones in 1835 at a scientific meeting but never removed his biblical blinders. He rejected Schmerling's theory and, instead of saying so quietly, proceeded to humiliate him. Buckland often claimed that because of various chemical changes, fossilized bones naturally stick to the tongue, whereas fresher bones don't. During a lecture at the meeting, Buckland placed onto his tongue one of the animal bones (a bear's) that Schmerling had found mixed in with the hominid remains. The bear bone stuck fast, and Buckland continued to lecture, the bone flopping about hilariously. He then challenged Schmerling to stick his "extinct human" bones to his own tongue. They fell off. Ergo they weren't ancient.

Though hardly definitive proof, the dismissal lingered in the minds of paleontologists. So when more uncanny skulls turned up in Gibraltar in 1848, prudent scientists ignored them. Eight years later—and just months after the death of Buckland, the last great Deluge scientist—miners worked loose still more of the odd bones from a limestone quarry in Germany's Neander Valley. One scholar, channeling Buckland, identified them as belonging to a deformed Cossack who had been injured by Napoleon's army and crawled into a cliff-side cave to die. But this time two other scientists reasserted that the remains belonged to a distinct line of hominids, a race more outcast than the biblical Ishmaelites. Perhaps it helped that, among the various bones, the duo had located an adult skullcap down to the eye sockets, which emphasized the thick, glowering brow we still associate with Neanderthals.*

With their eyes opened—and with the publication in 1859 of a little book by Charles Darwin—paleontologists began to find Neanderthals and related hominids across Africa, the Middle East, and Europe. The existence of ancient humans became a scientific fact. But just as predictably, the new evidence provoked new confusion. Skeletons can shift in the ground as rock formations buckle, fouling up attempts to date or interpret them. Bones also scatter and get crushed into smithereens, forcing scientists to rebuild entire creatures from a few stray molars or metatarsals—a subjective process open to dissension and differing interpretations. There's no guarantee either that scientists will find representative samples: if scientists in AD 1,000,000 discovered what remains of Wilt Chamberlain, Tom Thumb, and Joseph Merrick, would they even classify them as the same species? For these reasons, every new discovery of *Homo* this and *Homo* that in the 1800s and 1900s incited further and often nasty debate. And decade after decade passed without the ultimate questions (Were all archaic humanoids our ancestors? If not, how many twigs of humanity existed?) becoming clearer. As the old joke went, put twenty paleontologists into a room, and you'd get twenty-one different schemes for human evolution. One world expert in archaic human genetics, Svante Pääbo, has noted, "I'm often rather surprised about how much scientists fight in paleontology....I suppose the reason is that paleontology is a rather data-poor science. There are probably more paleontologists than there are important fossils in the world."

Such was the general state of things when genetics invaded paleontology and archaeology beginning in the 1960s—and *invaded* is the apt word. Despite their quarrels, U-turns, and antiquated tools, paleontologists and archaeologists had figured out a lot about human origins. They didn't need a savior, thanks. So many of them resented the intrusion of biologists with their DNA clocks and molecular-based family trees, hotshots intent on overturning decades of research with a single paper. (One anthropologist scoffed at the strictly molecular approach as "no muss, no fuss, no dishpan hands. Just throw some proteins into a laboratory apparatus, shake them up, and bingo!—we have an answer to questions that have puzzled us for three generations.") And really, the older scientists' skepticism was warranted: paleogenetics turned out to be beastly hard, and despite their promising ideas, paleogeneticists had to spend years proving their worth.

One problem with paleogenetics is that DNA is thermodynamically unstable. Over time, C chemically degrades into T, and G degrades into A, so paleogeneticists can't always believe what they read in ancient samples. What's more, even in the coldest climates, DNA breaks down into gibberish after 100,000 years; samples older than that harbor virtually no intact DNA. Even in relatively fresh samples, scientists might find themselves piecing together a billion-base-pair genome from fragments just fifty letters long—proportionally equivalent to reconstructing your typical hardcover from strokes, loops, serifs, and other fragments smaller than the tittle on an *i*.

Oh, and most of those fragments are junk. No matter where a corpse falls—the coldest polar ice cap, the driest Saharan dune—bacteria and fungi will worm inside and smear their own DNA around. Some ancient bones contain more than 99 percent foreign DNA, all of which must be laboriously extracted. And that's the easy kind of contamination to deal with. DNA spreads so easily from human contact (even touching or breathing on a sample can pollute it), and ancient hominid DNA so closely mirrors our own, that ruling out human contamination in samples is almost impossible.

These obstacles (plus a few embarrassing retractions over the years) have pushed paleogeneticists into near paranoia about contamination, and they demand controls and safeguards that seem a better fit for a biological warfare lab. Paleogeneticists prefer samples no human has ever handled—ideally, ones still dirty from remote dig sites, where workers use surgical masks and gloves and drop everything into sterile bags. Hair is the best material, since it absorbs fewer contaminants and can be bleached clean, but paleogeneticists will settle for less fragile bone. (And given the paucity of uncontaminated sites, they often settle for bones in museum storage lockers, especially bones so boring no one ever bothered studying them before.)

The sample selected, scientists bring it into a "clean room" maintained at higher-than-normal air pressure, so that air currents—and more to the point, the floating scraps of DNA that can ride around on air currents—cannot flow inward when the door opens. Anyone allowed inside the room dresses toe to top in sterile scrubs with face masks and booties and two pairs of gloves, and they get pretty used to the odor of the bleach swabbed over most surfaces. (One lab bragged that its technicians, presumably while in their suits, are given sponge baths in bleach.) If the sample is bone, scientists use dentist drills or picks to shave off a few grams of powder. They might even doctor the drill so it rotates only at 100 rpm, since the heat of a standard, 1,000-rpm drill can fry DNA. They then dissolve the nib of powder with chemicals, which liberates the DNA. At this point paleogeneticists often add tags—snippets of artificial DNA—to every fragment. That way they can tell if extraneous DNA, which lacks the tag,* ever infiltrates the sample after it leaves the clean room. Scientists might also note the racial backgrounds of technicians and other scientists (and probably even janitors) at the lab, so that if unexpected ethnic sequences show up, they can judge whether their sample was compromised.

After all this preparation, the actual DNA sequencing begins. We'll talk about this process in detail later, but basically scientists determine the A-C-G-T sequence of each individual DNA fragment, then use sophisticated software to piece the many, many fragments together. Paleogeneticists have successfully applied this technique to stuffed quaggas, cave bear skulls, woolly mammoth tufts, bees in amber, mummy skin, even Buckland's beloved coprolites. But the most spectacular work along these lines comes from Neanderthal DNA. After the discovery of Neanderthals, many scientists classified them as archaic humans—the first (before the metaphor became tired) missing link. Others put Neanderthals on their own terminal branch of evolution, while some European scientists considered Neanderthals the ancestors of some human races but not others. (Again, *sigh*, you can guess which races they singled out, Africans and Aborigines.) Regardless of the exact taxonomy, scientists considered Neanderthals thick-witted and lowbrow, and it didn't surprise anyone that they'd died out. Eventually some dissenters began arguing that Neanderthals showed more smarts than they got credit for: they used stone tools, mastered fire, buried their dead (sometimes with wildflowers), cared for the weak and lame, and possibly wore jewelry and played bone flutes. But the scientists couldn't prove that Neanderthals hadn't watched humans do these things first and aped them, which hardly takes supreme intelligence.

DNA, though, permanently changed our view of Neander-thals. As early as 1987, mitochondrial DNA showed that Nean-derthals weren't direct human ancestors. At the same time, when the complete Neanderthal genome appeared in 2010, it turned out that the butt of so many *Far Side* cartoons was pretty darn human anyway; we share well north of 99 percent of our genome with them. In some cases this overlap was homely: Neanderthals likely had reddish hair and pale skin; they had the most common blood type worldwide, O; and like most humans, they couldn't digest milk as adults. Other findings were more profound. Nean-derthals had similar MHC immunity genes, and also shared a gene, *foxp2*, associated with language skills, which means they may have been articulate.

It's not clear yet whether Neanderthals had alternative ver-sions of *apoE*, but they got more of their protein from flesh than we did and so probably had some genetic adaptations to metabo-lize cholesterol and fight infections. Indeed, archaeological evi-dence suggests that Neanderthals didn't hesitate to eat even their own dead—perhaps as part of primitive shamanic rituals, per-haps for darker reasons. At a cave in northern Spain, scientists have discovered the fifty-thousand-year-old remains of twelve murdered Neanderthal adults and children, many of them related. After the deed, their probably starving assailants butchered them with stone tools and cracked their bones to suck the mar-row, cannibalizing every edible ounce. A gruesome scene, but it was from this surfeit of 1,700 bones that scientists extracted much of their early knowledge of Neanderthal DNA.

Like it or not, similar evidence exists for human cannibal-ism. Each hundred-pound adult, after all, could provide starving comrades with forty pounds of precious muscle protein, plus edible fat, gristle, liver, and blood. More uncomfortably, archaeo-logical evidence has long suggested that humans tucked into each other even when not famished. But for years questions persisted about whether most nonstarvation cannibalism was religiously motivated and selective or culinary and routine. DNA suggests routine. Every known ethnic group worldwide has one of two genetic signatures that help our bodies fight off certain diseases that cannibals catch, especially mad-cow-like diseases that come from eating each other's brains. This defensive DNA almost certainly wouldn't have become fixed worldwide if it hadn't once been all too necessary.

As the cannibalism DNA shows, scientists don't rely entirely on ancient artifacts for information about our past. Modern human DNA holds clues as well. And about the first thing scientists noticed when they began surveying modern human DNA is its lack of variety. Roughly 150,000 chimps and around the same number of gorillas are living today, compared to some seven billion humans. Yet humans have less genetic diversity than these monkeys, significantly less. This suggests that the worldwide population of humans has dipped far below the population of chimps and gorillas recently, perhaps multiple times. Had the Endangered Species Act existed way back when, *Homo sapiens* might have been the Paleolithic equivalent of pandas and condors.

Scientists disagree on why our population decreased so much, but the origins of the debate trace back to two different theories—or really, two different weltanschauungs—first articulated in William Buckland's day. Virtually every scientist before then upheld a catastrophist view of history—that floods, earthquakes, and other cataclysms had sculpted the planet quickly, throwing up mountains over a long weekend and wiping out species overnight. A younger generation—especially Buckland's student Charles Lyell, a geologist—pushed gradualism, the idea that winds, tides, erosion, and other gentle forces shaped the earth and its inhabitants achingly slowly. For various reasons (including some posthumous smear campaigns), gradualism became associated with proper science, catastrophism with lazy reasoning and theatrical biblical miracles, and by the early 1900s catastrophism itself had been (and this puts it mildly) annihilated in science. Eventually the pendulum swung back, and catastrophism became respectable again after 1979, when geologists discovered that a city-sized asteroid or comet helped eradicate the dinosaurs. Since then, scientists have accepted that they can uphold a proper gradualist view for most of history and still allow that some pretty apocalyptic events have taken place. But this acceptance makes it all the more curious that one ancient calamity, the first traces of which were discovered within a year of the dino impact, has received far less attention. Especially considering that some scientists argue that the Toba supervolcano almost eliminated a species far more dear to us than dinosaurs: *Homo sapiens*.

Getting a grasp on Toba takes some imagination. Toba is—or was, before the top 650 cubic miles blew off—a mountain in Indonesia that erupted seventy-odd thousand years ago. But because no witnesses survived, we can best appreciate its terror by comparing it (however faintly) to the second-largest known eruption in that archipelago, the Tambora eruption of 1815.

In early April 1815, three pillars of fire straight out of Exodus blasted out of Tambora's top. Tens of thousands died as psychedelic orange lava surfed down the mountainside, and a tsunami five feet high and traveling at 150 miles per hour battered nearby islands. People fifteen hundred miles away (roughly from New York to mid–South Dakota) heard the initial blast, and the world went black for hundreds of miles around as a smoke plume climbed ten miles into the sky. That smoke carried with it enormous amounts of sulfurous chemicals. At first, these aerosols seemed harmless, even pleasant: in England, they intensified the pinks, oranges, and bloody reds of the sunsets that summer, a celestial drama that probably influenced the land- and sunscapes of painter J. M. W. Turner. Later effects were less cute. By 1816—popularly known as The Year Without a Summer—the sulfurous ejecta had mixed homogeneously into the upper atmosphere and began reflecting sunlight back into space. This loss of heat caused freak July and August snowstorms in the fledgling United States, and crops failed widely (including Thomas Jefferson's corn at Monticello). In Europe Lord Byron wrote a dire poem in July 1816 called "Darkness," which opens, "I had a dream, which was not all a dream. / The bright sun was extinguish'd... / Morn came and went—and came, and brought no day, / And men... / Were chill'd into a selfish prayer for light." A few writers happened to holiday with Byron that summer near Lake Geneva, but they found the days so dreary that they mostly sulked indoors. Channeling their mood, some took to telling ghost stories for entertainment—one of which, by young Mary Shelley, became *Frankenstein*.

Now, with all that in mind about Tambora, consider that Toba spewed for five times longer and ejected a dozen times more material—millions of tons of vaporized rock per second* at its peak. And being so much bigger, Toba's enormous black basilisk of a plume could do proportionately more damage. Because of prevailing winds, most of the plume drifted westward. And some scientists think that a DNA bottleneck began when the smoke, after sweeping across south Asia, scythed into the very grasslands in Africa where humans lived. According to this theory, the destruction happened in two phases. In the short term, Toba dimmed the sun for six years, disrupted seasonal rains, choked off streams, and scattered whole cubic miles of hot ash (imagine wading through a giant ashtray) across acres and acres of plants, a major food source. It's not hard to imagine the human population plummeting. Other primates might have suffered less at first because humans camped on the eastern edge of Africa, in Toba's path, whereas most primates lived inland, sheltered somewhat behind mountains. But even if Toba spared other animals initially, no one escaped the second phase. Earth was already mired in an Ice Age in 70,000 BC, and the persistent reflection of sunlight into space might well have exacerbated it. We have evidence that the average temperature dropped twenty-plus degrees in some spots, after which the African savannas—our ancient homes—probably contracted like puddles in August heat. Overall, then, the Toba-bottleneck theory argues that the initial eruption led to widespread starvation, but the deepening Ice Age is what really pinned the human population down.

Macaque, orangutan, tiger, gorilla, and chimpanzee DNA also show some signs of bottlenecking right around Toba, but humans really suffered. One study suggested that the human population, worldwide, might have dropped to forty adults. (The world record for fitting people in a phone booth is twenty-five.) That's an outlandishly pessimistic guess even among disaster scientists, but it's common to find estimates of a few thousand adults, below what some minor-league baseball teams draw. Given that these humans might not have been united in one place either, but scattered in small, isolated pockets around Africa, things look even shakier for our future. If the Toba-bottleneck theory is true, then the lack of diversity in human DNA has a simple explanation. We damn near went extinct.

Not surprisingly—more infighting—many archaeologists find that explanation for low genetic diversity way too pat, and the theory remains contentious. It's not the existence of a bottleneck per se that rankles. It's well established that the protohuman breeding population (roughly equivalent to the number of fertile adults) dropped alarmingly at times in the past million years. (Which, among other things, probably allowed a freakish trait like forty-six chromosomes to spread.) And many scientists see strong evidence in our DNA for at least one major bottleneck after anatomically modern humans arose 200,000 years ago.

What rankles scientists is linking any bottleneck to Toba; suspicion of the old, bad catastrophism looms.

Some geologists contest that Toba wasn't as powerful as their colleagues claim. Others doubt Toba could have decimated populations thousands of miles away, or that one puny mountain could throw up enough sulfurous spume to intensify a global ice age. Some archaeologists have also found evidence (disputed, inevitably) of stone tools right above and below some six-inch-thick Toba ash layers, which implies not extinction but continuity right where Toba should have done the most damage. We have genetic reasons to question a Toba bottleneck as well. Most important, geneticists simply cannot distinguish, retroactively, between the lack of diversity induced by a short but severe bottleneck and the lack of diversity induced by a longer but milder bottleneck. In other words, there's ambiguity: if Toba did crush us down to a few dozen adults, we'd see certain patterns in our DNA; but if a population was held down to a few thousand, as long as it was held down consistently, those people's DNA would show *the same* signatures after maybe a thousand years. And the wider the time frame, the less likely Toba had anything to do with the bottleneck.

William Buckland and others would have recognized this debate instantly: whether small but persistent pressures could have held our clever species down for so long, or whether it took a cataclysm. But it's a measure of progress that, unlike the rout of catastrophism in Buckland's day and the century of scorn that followed, modern scientific catastrophists can make their case heard. And who knows? The Toba supervolcano may yet join the dinosaur-killing space rock as one of the world's premier disasters.

So what does all this DNA archaeology add up to? As the field has come into its own, scientists have pulled together an overarching summary of how modern humans emerged and spread across the globe.

Perhaps most important, DNA confirmed our African origins. A few archaeological holdouts had always maintained that humankind arose in India or Asia, but a species generally shows the highest genetic diversity near its origin, where it's had the longest time to develop. That's exactly what scientists see in Africa. As one example, African peoples have twenty-two versions of a particular stretch of DNA linked to the all-important insulin gene—only three of which, total, appear in the rest of the world. For ages anthropologists lumped all African peoples into one "race," but the genetic truth is that the larger world's diversity is more or less a subset of African diversity.

DNA can also embellish the story of human origins with details about how we behaved long ago and even what we looked like. Around 220,000 years ago, the *apoE* meat-eating gene emerged and began to spread, introducing the possibility of productive old age. Just twenty thousand years later, another mutation allowed the hair on our heads to (unlike monkey hair or body hair) grow indefinitely long—a "haircut gene." Then, thirty thousand years later, we began wearing skins as clothes, a fact scientists determined by comparing the DNA clocks of head lice (which live only on scalps) and of related but distinct body lice (which live only in clothes) and figuring out when they diverged. In ways large and small, these changes transformed societies.

Suitably attired and perfectly coiffed, humans seem to have started making inroads from Africa into the Middle East maybe 130,000 years ago (our first imperial impulse). But something—cold weather, homesickness, predators, a Neanderthal NO TRESPASSING sign—halted their sprawl and drove them back to Africa. The human population bottlenecked over the next few tens of thousands of years, perhaps because of Toba. Regardless, humans scraped through and eventually recovered. But this time, instead of cowering and waiting for the next extinction threat, small clans of humans, as few as a few thousand people total, began establishing settlements beyond Africa, in waves beginning about sixty thousand years ago. These clans probably crossed the Red Sea at low tide, Moses style, through a southern point called Bab el Mandeb, the Gate of Grief. Because bottlenecks had isolated these clans for millennia, they'd developed unique genetic traits. So as they spread into new lands and their population doubled and redoubled, these traits blossomed into the unique features of European and Asian populations today. (In a touch Buckland would have appreciated, this multipronged dispersal from Africa is sometimes called the Weak Garden of Eden theory. But this tale is actually better than the biblical version; we didn't lose Eden but learned to make other Edens across the world.)

As we expanded beyond Africa, DNA kept a marvelous travelogue. In Asia, genetic analysis has revealed two distinct waves of human colonization: an earlier wave sixty-five thousand years ago that skirted India and led to the settlement of Australia, making the Aborigines the first real explorers in history; and a later wave that produced modern Asians and led to humanity's first population boom, forty thousand years ago, when 60 percent of humankind lived on the Indian, Malay, and Thai peninsulas. In North America, a survey of different genetic pools suggests that the first Americans paused for perhaps ten thousand years on the Bering land bridge between Siberia and Alaska, as if trembling to break away from Asia and enter the New World. In South America, scientists have discovered MHC genes from Amerindians in native Easter Islanders, and the thorough blending of those genes in the islanders' otherwise Asian chromosomes indicates that someone was making *Kon-Tiki*–like sea voyages back and forth to the Americas in the early 1000s, back when Columbus was mere specks of DNA spread among his great-great-great-(...)-great-grandparents' gonads. (Genetic analysis of sweet potatoes, bottle gourds, and chicken bones also indicates pre-Columbian contact.) In Oceania, scientists have tied the diffusion and winnowing of people's DNA to the diffusion and winnowing of their languages. It turns out that people in southern Africa, the cradle of humanity, not only have richer DNA than anyone else but richer languages, with up to one hundred distinct sounds, including the famous *tchk-tchk* clicks. Languages from intermediately diverse lands have fewer sounds (English has forty-some). Languages at the far end of our ancient migration, like Hawaiian, use around a dozen sounds, and Hawaiian people show correspondingly uniform DNA. It all adds up.

Looking beyond our species a tad, DNA can also illuminate one of the biggest mysteries of archaeology: what happened to Neanderthals? After Neanderthals had thrived in Europe for ages, something slowly strangled them into smaller and smaller territories, and the last ones expired about thirty-five thousand years ago in southern Europe. The profusion of theories to explain what doomed them—climate change, catching diseases from humans, competition for food, homicide (by *Homo sapiens*), "mad Neanderthal" disease from eating too many brains—is an unmistakable sign that nobody has the faintest. But with the decoding of the Neanderthal genome, we know at last that Neanderthals didn't disappear, not entirely. We carried their seed inside us across the globe.

After emerging from Africa around sixty thousand years ago, clans of humans eventually wandered into Neanderthal lands in the Levantine. Boys eyed girls, tyrannical hormones took over, and pretty soon little humanderthals were running around—a replay of when archaic humanoids bedded proto-chimpanzees (*plus ça change*). What happened next is hazy, but the groups divorced and did so asymmetrically. Perhaps some

outraged human elders stormed off, taking their despoiled children and humanderthal grandchildren with them. Perhaps only
Neanderthal men bagged human women, who then left with
their clan. Perhaps the groups parted amicably, but all the half-
breeds left in Neanderthal care died when the humans pushed
on and colonized the planet at large. Anyhow, when these Paleolithic Lewises and Clarks parted from their Neanderthal lovers,
they carried some Neanderthal DNA in their gene pool. Enough,
in fact, that we all still have a few percent's worth inside us
today—equivalent to the amount you inherited from each great-
great-great-grandparent. It's not yet clear what all this DNA
does, but some of it was MHC immunity DNA—which means
that Neanderthals may have unwittingly helped destroy themselves by giving humans the DNA to fight new diseases in the
new lands we took from Neanderthals. Oddly, however, there
seems to have been no reciprocity: no uniquely human DNA,
disease-fighting or otherwise, has shown up in any Neanderthals so far. No one knows why.

And actually only some of us absorbed Neanderthal DNA.
All the amour took place on the pivot between Asia and Europe,
not in Africa proper. Which means that the people who carried
Neanderthal DNA forward weren't ancient Africans (who, inso-
far as scientists can tell, never hooked up with Neanderthals) but
the early Asians and Europeans, whose descendants populated
the rest of the world. The irony is too rich not to point out.
When arranging the different human races in tiers, from just
below the angels to just above the brutes, smug racialist scientists of the 1800s always equated black skin with "subhuman"
beasts like Neanderthals. But facts is facts: pure Nordic Europeans carry far more Neanderthal DNA than any modern African.
One more time, DNA debases.

Just to frustrate archaeologists, however, evidence turned up
in 2011 that Africans had their own extra-species liaisons. Cer-

tain tribes who stayed at home in central Africa and never saw a Neanderthal in their lives seem to have acquired chunks of non-coding DNA from other, unnamed, and now-extinct archaic humans, and did so well after the early Asians and Europeans had left. As scientists continue to catalog human diversity around the world, DNA memories of other assignations will no doubt surface in other groups, and we'll have to ascribe more and more "human" DNA to other creatures.

Really, though, tallying whether this or that ethnic group has less archaic DNA than another misses the point. The emerging and vital truth isn't who is more Neanderthal than whom. It's that all peoples, everywhere, enjoyed archaic human lovers whenever they could. These DNA memories are buried deeper inside us than even our ids, and they remind us that the grand saga of how humans spread across the globe will need some personal, private, all-too-human amendments and annotations—rendezvous here, elopements there, and the commingling of genes most everywhere. At least we can say that all humans are united in sharing this shame (if shame it is) and in sharing these scarlet A's, C's, G's, and T's.

11

Size Matters

How Did Humans Get Such Grotesquely Large Brains?

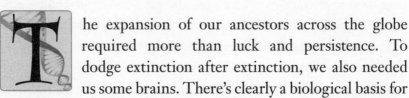he expansion of our ancestors across the globe required more than luck and persistence. To dodge extinction after extinction, we also needed us some brains. There's clearly a biological basis for human intelligence; it's too universal not to be inscribed in our DNA, and (unlike most cells) brain cells use almost all the DNA we have. But despite centuries of inquiry, by everyone from phrenologists to NASA engineers, on subjects from Albert Einstein to idiot savants, no one quite knows where our smarts come from.

Early attempts to find the biological basis of intelligence played off the idea that bigger was better: more brain mass meant more thinking power, just like more muscles meant more lifting power. Although intuitive, this theory has its shortcomings; whales and their twenty-pound brains don't dominate the globe. So Baron Cuvier, the half Darwin, half Machiavelli from Napoleonic France, suggested that scientists also examine a creature's brain-body ratio, to measure its relative brain weight as well.

Nonetheless scientists in Cuvier's day maintained that bigger brains did mean finer minds, especially *within* a species. The best evidence here was Cuvier himself, a man renowned (indeed, practically stared at) for the veritable pumpkin atop his shoulders. Still, no one could say anything definitive about Cuvier's brain until 7 a.m. on Tuesday, May 15, 1832, when the greatest and most shameless doctors in Paris gathered to conduct Cuvier's autopsy. They sliced open his torso, sluiced through his viscera, and established that he had normal organs. This duty dispatched, they eagerly sawed through his skull and extracted a whale of a specimen, sixty-five ounces, over 10 percent larger than any brain measured before. The smartest scientist these men had ever known had the biggest brain they'd ever seen. Pretty convincing.

By the 1860s, though, the tidy size-smarts theory had started unraveling. For one, some scientists questioned the accuracy of the Cuvier measurement—it just seemed too outré. No one had bothered to pickle and preserve Cuvier's brain, unfortunately, so these later scientists grasped at whatever evidence they could find. Someone eventually dug up Cuvier's hat, which was indeed commodious; it fell over the eyes of most everyone who donned it. But those wise in the ways of milliners pointed out that the hat's felt might have stretched over the years, leading to overestimates. Tonsorial types suggested instead that Cuvier's bushy hairdo had made his head merely appear enormous, biasing his doctors to expect (and, because expecting, find) a vast brain. Still others built a case that Cuvier suffered from juvenile hydrocephaly, a feverish swelling of the brain and skull when young. In that case, Cuvier's big head might be accidental, unrelated to his genius.*

Arguing about Cuvier wasn't going to solve anything, so to get more data on more people, cranial anatomists developed methods to gauge the volumes of skulls. Basically, they plugged every hole and filled the skulls (depending on their preference)

Baron Cuvier—a half-Darwin, half-Machiavelli biologist
who lorded over French science during and after Napoleon—had one of the largest human brains ever recorded.
(James Thomson)

with a known amount of peas, beans, rice, millet, white pepper-
corn, mustard seed, water, mercury, or lead buckshot. Imagine
rows of skulls on a table, funnels protruding from each one, an
assistant humping around buckets of quicksilver or burlap bags
of grain from the market. Whole monographs were published
on these experiments, but they produced still more baffling
results. Were Eskimos, who had the largest brainpans, really the
smartest people on earth? What's more, the skulls of the newly
discovered Neanderthal species were actually roomier than human
skulls by an average of six cubic inches.

As it turns out, that's just the start of the confusion. Again, without correlating strictly, a bigger brain does generally make a species smarter. And because monkeys, apes, and humans are all pretty sharp, scientists assumed that intense pressure must have come to bear on primate DNA to boost brain size. It was basically an arms race: big-brain primates win the most food and survive crises better, and the only way to beat them is to get smarter yourself. But nature can be stingy, too. Based on genetic and fossil evidence, scientists can now track how most primate lineages have evolved over many millions of years. It turns out that certain species' bodies, and not infrequently their brains, shrank over time—they became cranial runts. Brains consume a lot of energy (around 20 percent of human calories), and in times with chronic food shortages, the DNA that won out in primates was the miserly DNA that scrimped on building up brains.

The best-known runt today is probably the "hobbit" skeleton from the Indonesian island of Flores. When it was discovered in 2003, many scientists declared it a stunted or microcephalic (tiny-headed) human; no way evolution was irresponsible enough to let the brains of a hominid dwindle that much, brains being about all we hominids have going. But nowadays most scientists accept that the brains of hobbits (officially, *Homo floresiensis*) did shrink. Some of this diminution might relate to so-called island dwarfism: islands, being severely finite, have less food, so if an animal can tune down some of the hundreds of genes that control its height and size, it can get by with fewer calories. Island dwarfism has shrunk mammoths, hippos, and other stranded species to pygmy sizes, and there's no reason to think this pressure wouldn't squash a hominid, even if the cost is a punier brain.*

By some measures, modern humans are runts, too. We've probably all gone to a museum and snickered over the wee suit of armor that a king of England or some other big swinging dick from history wore—what a shrimp! But our ancestors would giggle at our clothes just the same. Since about 30,000 BC, our DNA has diminished the average human body size by 10 percent (roughly five inches). The vaunted human brain dwindled by at least 10 percent over that span, too, and a few scientists argue it has shrunk even more.

Scientists filling skulls with buckshot or millet in the early 1900s didn't know about DNA, of course, but even with their crude tools, they could tell the brain size–intelligence theory didn't add up. One famous study of geniuses—it got a two-page spread in the *New York Times* in 1912—did find some truly capacious organs. Russian writer Ivan Turgenev's brain topped seventy ounces, compared to a human average of fifty. At the same time, the brains of statesman Daniel Webster and mathematician Charles Babbage, who dreamed up the first programmable computer, were merely average. And poor Walt Whitman had to sound his barbaric yawp over the rooftops with a command center of just forty-four ounces. Even worse was Franz Joseph Gall. Though an intelligent scientist—he proposed for the first time that different brain regions have different functions—Gall also founded phrenology, the analysis of head lumps. To his followers' eternal shame, he weighed in at a measly forty-two ounces.

To be fair, a technician dropped Whitman's brain before measuring it. It crumbled into pieces like a dried-out cake, and it's not clear whether they found all of them, so maybe Walt could have made a better showing. (No such mishap with Gall.) Regardless, by the 1950s the size-smarts theory had received some fatal wounds, and any lingering association between brain heft and braininess died for good a few hours after Albert Einstein himself died in 1955.

After suffering an aortic aneurysm on April 13, 1955, Einstein found himself the subject of an international death watch. He finally succumbed to internal hemorrhaging at 1:15 a.m. on April 18. His body arrived shortly thereafter at a local hospital in Princeton, New Jersey, for a routine autopsy. At this point the pathologist on duty, Thomas Harvey, faced a stark choice.

Any one of us might have been tempted the same way—who wouldn't want to know what made Einstein *Einstein?* Einstein himself expressed interest in having his brain studied after he died, and even sat for brain scans. He decided against preserving the best part of himself only because he loathed the thought of people venerating it, the twentieth-century equivalent of a medieval Catholic relic. But as Harvey arranged the scalpels in his autopsy room that night, he knew humankind had just one chance to salvage the gray matter of the greatest scientific thinker in centuries. And while it may be too strong to say *stole*, by 8 a.m. the next morning—without next-of-kin permission, and against Einstein's notarized wish for cremation—Harvey had shall we say *liberated* the physicist's brain and released the body to the family without it.

The disappointment started immediately. Einstein's brain weighed forty-three ounces, at the low end of normal. And before Harvey could measure anything more, word of the relic spread, just as Einstein had feared. During a discussion in school the next day about the loss of Einstein, Harvey's son, normally a laconic lad, blurted out, "My dad's got his brain!" A day later, newspapers across the country mentioned Harvey's plans in their front-page obits. Harvey did eventually convince the remaining Einsteins, who were sure peeved, to grant permission for further study. So after measuring its dimensions with calipers and photographing it for posterity with his 35 mm black-and-white camera, Harvey sawed the brain into 240 taffy-sized hunks and lacquered each one in celloidin. Harvey was soon mailing the blobs in mayo jars to neurologists, confident that the forthcoming scientific insights would justify his peccadillo.

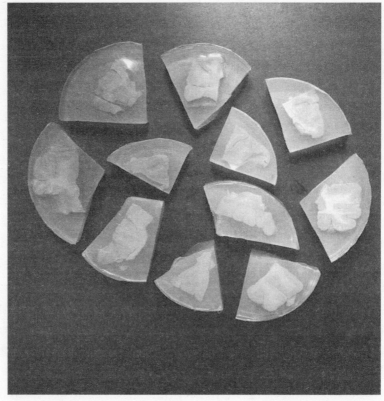

Fragments of Einstein's brain, shellacked in hard celloidin after the physicist's death in 1955. (Getty Images)

This certainly wasn't the first autopsy of a famous person to take a lurid turn. Doctors set aside Beethoven's ear bones in 1827 to study his deafness, but a medical orderly nicked them. The Soviet Union founded an entire institute in part to study Lenin's brain and determine what makes a revolutionary a revolutionary. (The brains of Stalin and Tchaikovsky also merited preservation.) Similarly, and despite the body being mutilated by mobs, Americans helped themselves to half of Mussolini's brain after World War II, to determine what made a dictator a dictator. That same year the U.S. military seized four thousand pieces of human flesh from Japanese coroners to study nuclear radiation damage. The spoils included hearts, slabs of liver and brain, even disembodied eyeballs, all of which doctors stored in jars in radiation-proof vaults in Washington, D.C., at a cost to taxpayers of $60,000 per year. (The U.S. repatriated the remains in 1973.)

Even more grotesquely, William Buckland—in a story that's possibly apocryphal, but that his contemporaries believed—topped his career as a gourmand when a friend opened a silver snuffbox to show off a desiccated morsel of Louis XIV's heart. "I have eaten many strange things, but have never eaten the heart of a king," Buckland mused. Before anyone thought to stop him, Buckland wolfed it down. One of the all-time raciest stolen body parts was the most private part of Cuvier's patron, Napoleon. A spiteful doctor lopped off L'Empereur's penis during the autopsy in 1821, and a crooked priest smuggled it to Europe. A century later, in 1927, the unit went on sale in New York, where one observer compared it to a "maltreated strip of buckskin shoelace." It had shriveled to one and one-half inches, but a urologist in New Jersey bought it anyway for $2,900. And we can't wrap up this creepy catalog without noting that yet another New Jersey doctor disgracefully whisked away Einstein's eyeballs in 1955. The doctor later refused Michael Jackson's offer to pay millions for them—partly because the doc had grown fond of gazing into them. As for the rest of Einstein's body, take heart (sorry). It was cremated, and no one knows where in Princeton his family scattered the ashes.*

Perhaps the most disheartening thing about the whole Einstein fiasco is the paltry knowledge scientists gained. Neurologists ended up publishing only three papers on Einstein's brain in forty years, because most found nothing extraordinary there. Harvey kept soliciting scientists to take another look, but after the initial null results came back, the brain chunks mostly just sat around. Harvey kept each section wrapped in cheesecloth and piled them into two wide-mouthed glass cookie jars full of formaldehyde broth. The jars themselves sat in a cardboard box labeled "Costa Cider" in Harvey's office, tucked behind a red beer cooler. When Harvey lost his job later and took off for greener pastures in Kansas (where he moved in next door to author and junkie William S. Burroughs), the brain rode shotgun in his car.

In the past fifteen years, though, Harvey's persistence has been justified, a little. A few cautious papers have highlighted some atypical aspects of Einstein's brain, on both microscopic and macroscopic levels. Coupled with loads of research into the genetics of brain growth, these findings may yet provide some insight into what separates a human brain from an animal brain, and what pushes an Einstein a few standard deviations beyond that.

First, the obsession with overall brain size has given way to obsessing over the size of certain brain parts. Primates have particularly beefy neuron shafts (called axons) compared to other animals and can therefore send information through each neuron more quickly. Even more important is the thickness of the cortex, the outermost brain layer, which promotes thinking and dreaming and other flowery pursuits. Scientists know that certain genes are crucial for growing a thick cortex, partly because it's so sadly obvious when these genes fail: people end up with primitively tiny brains. One such gene is *aspm*. Primates have extra stretches of DNA in *aspm* compared to other mammals, and this DNA codes for extra strings of amino acids that bulk up the cortex. (These strings usually start with the amino acids isoleucine and glutamine. In the alphabetic abbreviations that biochemists use for amino acids, glutamine is usually shortened to Q [G was taken] and isoleucine to plain I—which means we probably got an intelligence boost from a string of DNA referred to, coincidentally, as the "IQ domain.")

In tandem with increasing cortex size, *aspm* helps direct a process that increases the density of neurons in the cortex, another trait that correlates strongly with intelligence. This increase in density happens during our earliest days, when we have loads of stem cells, undeclared cells that can choose any path and become any type of cell. When stem cells begin dividing in the incipient brain, they can either produce more stem cells, or they can settle down, get a job, and become mature neurons. Neurons are good, obviously, but each time a neuron forms, the production of new stem cells (which can make additional neurons in the future) stops. So getting a big brain requires building up the base population of stem cells first. And the key to doing that is making sure that stem cells divide evenly: if the cellular guts get divided equally between both daughter cells, each one becomes another stem cell. If the split is unequal, neurons form prematurely.

To facilitate an even split, *aspm* guides the "spindles" that attach to chromosomes and pull them apart in a nice, clean, symmetrical way. If *aspm* fails, the split is uneven, neurons form too soon, and the child is cheated of a normal brain. To be sure, *aspm* isn't *the* gene responsible for big brains: cell division requires intricate coordination among many genes, with master regulator genes conducting everything from above, too. But *aspm* can certainly pack the cortex with neurons* when it's firing right—or sabotage neuron production if it misfires.

Einstein's cortex had a few unusual features. One study found that, compared to normal elderly men, his had the same number of neurons and the same average neuron size. However, part of Einstein's cortex, the prefrontal cortex, was thinner, which gave him a greater density of neurons. Closely packed neurons may help the brain process information more quickly—a tantalizing find considering that the prefrontal cortex orchestrates thoughts throughout the brain and helps solve multistep problems.

Further studies examined certain folds and grooves in Einstein's cortex. As with brain size, it's a myth that simply having more folds automatically makes a brain more potent. But folding does generally indicate higher functioning. Smaller and dumber monkeys, for instance, have fewer corrugations in their cortexes. As, interestingly, do newborn humans. Which means that as we mature from infants to young adults, and as genes that wrinkle our brains start kicking on, every one of us relives millions of years of human evolution. Scientists also know that a lack of brain folds is devastating. The genetic disorder "smooth brain" leaves babies severely retarded, if they even survive to term. Instead of being succulently furrowed, a smooth brain looks eerily polished, and cross sections of it, instead of showing scrunched-up brain fabric, look like slabs of liver.

Einstein had unusual wrinkles and ridges in the cortex of his parietal lobe, a region that aids in mathematical reasoning and image processing. This comports with Einstein's famous declaration that he thought about physics mostly through pictures: he formulated relativity theory, for instance, in part by imagining what would happen if he rode around bareback on light rays. The parietal lobe also integrates sound, sight, and other sensory input into the rest of the brain's thinking. Einstein once declared that abstract concepts achieved meaning in his mind "only through their connection with sense-experiences," and his family remembers him practicing his violin whenever he got stuck with a physics problem. An hour later, he'd often declare, "I've got it!" and return to work. Auditory input seemed to jog his thinking. Perhaps most telling, the parietal wrinkles and ridges in Einstein's lobes were steroid thick, 15 percent bigger than normal. And whereas most of us mental weaklings have skinny right parietal lobes and even skinnier left parietal lobes, Einstein's were equally buff.

Finally, Einstein appeared to be missing part of his middle brain, the parietal operculum; at the least, it didn't develop fully. This part of the brain helps produce language, and its lack might explain why Einstein didn't speak until age two and why until age seven he had to rehearse every sentence he spoke aloud under his breath. But there might have been compensations. This region normally contains a fissure, or small gap, and our thoughts get routed the long way around. The lack of a gap might have meant that Einstein could process certain information more speedily, by bringing two separate parts of his brain into unusually direct contact.

All of which is exciting. But is it exciting bunkum? Einstein feared his brain becoming a relic, but have we done something equally silly and reverted to phrenology? Einstein's brain has deteriorated into chopped liver by now (it's even the same color), which forces scientists to work mostly from old photographs, a less precise method. And not to put too fine a point on it, but Thomas Harvey coauthored half of the various studies on the "extraordinary" features of Einstein's brain, and he certainly had an interest in science learning something from the organ he purloined. Plus, as with Cuvier's swollen brain, maybe Einstein's features are idiosyncratic and had nothing to do with genius; it's hard to tell with a sample size of one. Even trickier, we can't sort out if unusual neurofeatures (like thickened folds) caused Einstein's genius, or if his genius allowed him to "exercise" and build up those parts of his brain. Some skeptical neuroscientists note that playing the violin from an early age (and Einstein started lessons at six) can cause the same brain alterations observed in Einstein.

And if you had hopes of dipping into Harvey's brain slices and extracting DNA, forget it. In 1998, Harvey, his jars, and a writer took a road trip in a rented Buick to visit Einstein's granddaughter in California. Although weirded out by Grandpa's brain, Evelyn Einstein accepted the visitors for one reason. She was poor, reputedly dim, and had trouble holding down a job — not exactly an Einstein. In fact Evelyn was always told she'd been adopted by Einstein's son, Hans. But Evelyn could do a little math, and when she started hearing rumors that Einstein had canoodled with various lady friends after his wife died, Evelyn realized she might be Einstein's bastard child. The "adoption" might have been a ruse. Evelyn wanted to do a genetic paternity test to settle things, but it turned out that the embalming process had denatured the brain's DNA. Other sources of his DNA might still be floating around—strands in mustache brushes, spittle on pipes, sweated-on violins—but for now we know more about the genes of Neanderthals who died fifty thousand years ago than the genes of a man who died in 1955.

But if Einstein's genius remains enigmatic, scientists have sussed out a lot about the everyday genius of humans compared to that of other primates. Some of the DNA that enhances human intelligence does so in roundabout ways. A two-letter frameshift mutation in humans a few million years ago deactivated a gene that bulked up our jaw muscles. This probably allowed us to get by with thinner, more gracile skulls, which in turn freed up precious cc's of skull for the brain to expand into. Another surprise was that *apoE*, the meat-eating gene, helped a lot, by helping the brain manage cholesterol. To function properly, the brain needs to sheathe its axons in myelin, which acts like rubber insulation on wires and prevents signals from short-circuiting or misfiring. Cholesterol is a major component of myelin, and certain forms of *apoE* do a better job distributing brain cholesterol where it's needed. *ApoE* also seems to promote brain plasticity.

Some genes lead to direct structural changes in the brain. The *lrrtm1* gene helps determine which exact patches of neurons control speech, emotion, and other mental qualities, which in turn helps the human brain establish its unusual asymmetry and left-right specialization. Some versions of *lrrtm1* even reverse parts of the left and right brain—and increase your chances of being left-handed to boot, the only known genetic association for that trait. Other DNA alters the brain's architecture in almost comical ways: certain inheritable mutations can cross-wire the sneeze reflex with other ancient reflexes, leaving people achooing uncontrollably—up to forty-three times in a row in one case—after looking into the sun, eating too much, or having an orgasm. Scientists have also recently detected 3,181 base pairs of brain "junk DNA" in chimpanzees that got deleted in humans. This region helps stop out-of-control neuron growth, which can lead to big brains, obviously, but also brain tumors. Humans gambled in deleting this DNA, but the risk apparently paid off, and our brains ballooned. The discovery shows that it's not always what we gained with DNA, but sometimes what we lost, that makes us human. (Or at least makes us nonmonkey: Neanderthals didn't have this DNA either.)

How and how quickly DNA spreads through a population can reveal which genes contribute to intelligence. In 2005 scientists reported that two mutated brain genes seem to have swept torrentially through our ancestors, *microcephalin* doing so 37,000 years ago, *aspm* just 6,000 years ago. Scientists clocked this spread by using techniques first developed in the Columbia fruit fly room. Thomas Hunt Morgan discovered that certain versions of genes get inherited in clusters, simply because they reside near each other on chromosomes. As an example, the A, B, and D versions of three genes might normally appear together; or (lowercase) a, b, and d might appear together. Over time, though, chromosomal crossing-over and recrossing will mix the groups, giving combos like a, B, and D; or A, b, and D. After enough generations, every combination will appear.

But say that B mutates to B_0 at some point, and that B_0 gives people a hell of a brain tune-up. At that point it could sweep through a population, since B_0 people can outthink everyone else. (That spread will be especially easy if the population drops very low, since the novel gene has less competition. Bottlenecks aren't always bad!) And notice that as B_0 sweeps through a population, the versions of A/a and D/d that happen to be sitting next to B_0 in the first person with the mutation will *also* sweep through the population, simply because crossing over won't have time to break the trio apart. In other words, these genes will ride along with the advantageous gene, a process called genetic hitchhiking. Scientists see especially strong signs of hitchhiking with *aspm* and *microcephalin*, which means they spread especially quickly and probably provided an especially strong advantage.

Beyond any specific brain-boosting genes, DNA regulation might explain a lot about our gray matter. One flagrant difference between human and monkey DNA is that our brain cells splice DNA far more often, chopping and editing the same string of letters for many different effects. Neurons mix it up so much, in fact, that some scientists think they've upended one central dogma of biology—that all cells in your body have the same DNA. For whatever reason, our neurons allow much more free play among mobile DNA bits, the "jumping genes" that wedge themselves randomly into chromosomes. This changes the DNA patterns in neurons, which can change how they work. As one neuroscientist observes, "Given that changing the firing patterns of single neurons can have marked effects on behavior...it is likely that some [mobile DNA], in some cells, in some humans, will have significant, if not profound, effects on the final structure and function of the human brain." Once again viruslike particles may prove important to our humanity.

If you're skeptical that we can explain something as ineffable as genius by studying something as reductive as DNA, a lot of scientists are right there with you. And every so often a case like that of savant Kim Peek pops up—a case that so mocks our understanding of how DNA and brain architecture influence intelligence that even the most enthusiastic neuroscientist seeks the consolation of a stiff bourbon and starts to think seriously about going into administration.

Peek, a Salt Lake City native, was actually a megasavant, a souped-up version of what's impolitely but accurately known as an idiot savant. Instead of being limited to poignantly empty skills like drawing perfect circles or listing all the Holy Roman Emperors in order, Peek had encyclopedic knowledge of geography, opera, American history, Shakespeare, classical music, the Bible—basically all of Western Civ. Even more intimidating, Peek had Google-like recall of any sentence in the nine thousand books he'd memorized, starting at eighteen months old. (When finished with a book, he returned it to his shelf with the spine upside down to indicate he'd knocked it off.) If it makes you less insecure, Peek did know loads of useless crap, too, like the complete U.S. zip code system. He also memorized *Rain Man*, a movie he inspired, and knew Mormon theology in lobotomizing detail.*

Trying to get some, any, sort of measure on Peek's talents, doctors in Utah began scanning his brain in 1988. In 2005, NASA got involved for whatever reason and took complete MRI and tomography scans of Peek's mental plumbing. The scans revealed that Peek lacked the tissue that connects the brain's right hemisphere to the left. (Peek's father remembered, in fact, that Peek could move each eye independently of the other as an infant, probably because of that disconnect between the right and left halves.) The left hemisphere, which focuses on big-picture ideas, also seemed misshaped—more lumpen and squished than normal brains. But beyond these details, scientists learned little. In the end, then, even NASA-level technology could reveal only abnormal features, *problems* with Peek's brain. If you wanted to know why Peek couldn't button his own clothes or why he could never remember where to find the silverware, despite living at his father's house for decades, there you go. As to the basis of his talents, NASA shrugged.

But doctors also knew Peek had a rare genetic disorder, FG syndrome. In FG syndrome, a single malfunctioning gene can't flick the on switch for a stretch of DNA that neurons need to develop properly. (They're very picky, neurons.) And as with most savants, the fallout from these problems clustered in Peek's left brain, possibly because the big-picture-oriented left hemisphere takes longer to develop in utero. A malfunctioning gene therefore has more time to inflict injury there. But in a strange twist, injuring the normally dominant left hemisphere can actually coax out the talents of the detail-conscious right brain. Indeed, the talents of most savants—artistic mimicry, perfect musical regurgitation, feats of calendary calculation—all cluster in the brain's less-vulnerable right half. Sadly, then, it may be true that those suppressed right-brained talents can never surface unless the domineering left hemisphere suffers damage.

Geneticists have made similar discoveries using the Neanderthal genome. Scientists are currently mining Neanderthal and human DNA for evidence of hitchhiking, trying to identify DNA that swept through humankind after the Neanderthal-human split and therefore helped distinguish us from Neanderthals. They've found about two hundred regions so far, most of which contain at least a few genes. Some of these human-Neanderthal differences are real yawners about bone development or metabolism. But scientists have also identified a handful of genes related to cognition. Paradoxically, though, having certain variants of these genes—far from being linked to Nobel Prizes or MacArthur grants—increases the risk of Down syndrome, autism, schizophrenia, and other mental disorders. It seems that a more complicated mind is a more fragile mind; if these genes did uplift our intelligence, taking them on also introduced risk.

And yet for all that frailty, a brain with faulty DNA can be miraculously resilient in other circumstances. In the 1980s, a neurologist in England scanned the freakishly large head of a young man referred to him for a checkup. He found little inside the skull case but cerebrospinal fluid (mostly salt water). The young man's cortex was basically a water balloon, a sac one millimeter thick surrounding a sloshing inner cavity. The scientist guessed the brain weighed perhaps five ounces. The young man also had an IQ of 126 and was an honors mathematics student at his university. Neurologists don't even pretend to know how these so-called high-functioning hydrocephalics (literally "water-heads") manage to live normal lives, but a doctor who studied another famous hydrocephalic, a civil servant in France with two children, suspects that if the brain atrophies slowly over time, it's plastic enough to reallocate important functions before it loses them completely.

Peek—who had a Cuvier-sized noggin himself—had an IQ of 87. It was probably so low because he reveled in minutiae and couldn't really process intangible ideas. For instance, scientists noticed he couldn't grasp common proverbs—the leap to the metaphorical was too far. Or when Peek's father told him to lower his voice once in a restaurant, Peek slid down in his chair, bringing his larynx closer to the ground. (He did seem to grasp that puns are theoretically funny, possibly because they involve a more mathematical substitution of meanings and words. He once responded to a query about Lincoln's Gettysburg Address by answering, "Will's house, 227 NW Front St. But he stayed there only one night—he gave the speech the next day.") Peek struggled with other abstractions as well and was basically helpless domestically, as reliant as a child on his father's care. But given his other talents, that 87 seems almost criminally unfair and certainly doesn't capture him.*

Peek died of a heart attack around Christmas in 2009, and

his body was buried. So there will be no Einstein-like afterlife for his remarkable brain. His brain scans still exist, but for now they mostly just taunt us, pointing out gaps in what we know about the sculpting of the human mind—what separated Peek from Einstein, or even what separates everyday human smarts from simian intelligence. Any deep appreciation of human intellect will require understanding the DNA that builds and designs the web of neurons that think our thoughts and capture each "Aha!" But it will also require understanding environmental influences that, like Einstein's violin lessons, goose our DNA and allow our big brains to fulfill their potential. Einstein was Einstein because of his genes, but not only because of them.

The environment that nurtured Einstein and the rest of us everyday geniuses didn't arise by accident. Unlike other animals, humans craft and design our immediate environments: we have culture. And while brain-boosting DNA was necessary for creating culture, it didn't suffice. We had big brains during our scavenger-gatherer days (perhaps bigger brains than now), but achieving sophisticated culture also required the spread of genes to digest cooked food and handle a more sedentary lifestyle. Perhaps above all we needed behavior-related genes: genes to help us tolerate strangers and live docilely under rulers and tolerate monogamous sex, genes that increased our discipline and allowed us to delay gratification and build things on generational time-scales. Overall, then, genes shaped what culture we would have, but culture bent back and shaped our DNA, too. And understanding the greatest achievements of culture—art, science, politics—requires understanding how DNA and culture intersect and evolve together.

12

The Art of the Gene

How Deep in Our DNA Is Artistic Genius?

rt, music, poetry, painting—there are no finer expressions of neural brilliance, and just like Einstein's or Peek's genius, genetics can illuminate some unexpected aspects of the fine arts. Genetics and visual art even trace a few parallel tracks across the past 150 years. Paul Cézanne and Henri Matisse couldn't have developed their arrestingly colorful styles if European chemists hadn't invented vibrant new dyes and pigments in the 1800s. Those dyes and pigments simultaneously allowed scientists to study chromosomes for the first time, because they could finally stain chromosomes a color different from the uniform blah of the rest of the cell. Chromosomes in fact take their name from the Greek for color, *chrōma*, and some techniques to tint chromosomes— like turning them "Congo red" on shimmering green backgrounds—would have turned Cézanne and Matisse a certain shade with envy. Meanwhile silver staining—a by-product of the new photographic arts—provided the first clean pictures of other cell structures, and photography itself allowed scientists to study time lapses of dividing cells and see how chromosomes got passed around.

Movements like cubism and Dadaism—not to mention competition from photography—led many artists to abandon realism and experiment with new kinds of art in the early twentieth century. And piggybacking on the insights gained from staining cells, photographer Edward Steichen introduced "bio-art" in the 1930s, with an early foray into genetic engineering. An avid gardener, Steichen began (for obscure reasons) soaking delphinium seeds in his gout medication one spring. This doubled the chromosome number in these purple flowers, and although some seeds produced "stunted, febrile rejects," others produced Jurassic-sized flora with eight-foot stalks. In 1936 Steichen exhibited five hundred delphiniums at the Museum of Modern Art in New York City and earned mostly rapturous reviews from newspapers in seventeen states: "Giant spikes...brilliant dark blues," one reviewer wrote, "a plum color never before seen...startlingly black eyes." The plums and blues may have startled, but Steichen—a nature-worshipping pantheist—echoed Barbara McClintock in insisting that the real art lay in controlling the development of the delphiniums. These views on art alienated some critics, but Steichen insisted, "A thing is beautiful if it fulfills its purpose—if it functions."

By the 1950s, a preoccupation with form and function eventually pushed artists into abstractionism. Studies of DNA coincidentally tagged along. Watson and Crick spent as many hours as any sculptor ever did creating physical models for their work, crafting various mock-ups of DNA from tin or cardboard. The duo settled on the double helix model partly because its austere beauty bewitched them. Watson once recalled that every time he saw a spiral staircase, he grew more convinced that DNA must look equally elegant. Crick turned to his wife, Odile, an artist, to draw the chic double helix that wound up and down the margin of their famous first paper on DNA. And later, Crick recalled a drunken Watson ogling their slender, curvy model one night and muttering, "It's so beautiful, you see, so beautiful." Crick added: "Of course, it was."

But like their guesses about the shapes of A, C, G, and T, Watson and Crick's guess about the overall shape of DNA rested on a somewhat shaky foundation. Based on how fast cells divide, biologists in the 1950s calculated that the double helix would have to unravel at 150 turns per second to keep up, a furious pace. More worrisome, a few mathematicians drew on knot theory to argue that separating the strands of helical DNA—the first step for copying them—was topologically impossible. That's because two unzipped helix strands cannot be pulled apart laterally—they're too intertwined, too entangled. So in 1976 a few scientists began promoting a rival "warped-zipper" structure for DNA. Here, instead of one long smooth right-handed helix, right- and left-handed half helixes alternated up and down the length of DNA, which would allow it to pull apart cleanly. To answer the criticisms of double helixes, Watson and Crick occasionally discussed alternative forms of DNA, but they (especially Crick) would almost immediately dismiss them. Crick often gave sound technical reasons for his doubts, but added once, tellingly, "Moreover, the models were ugly." In the end, the mathematicians proved right: cells cannot simply unwind double helixes. Instead they use special proteins to snip DNA, shake its windings loose, and solder it back together later. However elegant itself, a double helix leads to an awfully awkward method of replication.*

By the 1980s scientists had developed advanced genetic engineering tools, and artists began approaching scientists about collaborating on "genetic art." Honestly, your tolerance for bull-roar has to be pretty high to take some claims for genetic art seriously: pace bioartist George Gessert, do "ornamental plants, pets, sporting animals, and consciousness-altering drug plants" really constitute "a vast, unacknowledged genetic folk art"? And some perversities—like an albino bunny reconfigured with jelly-fish genes that made it glow green—were created, the artist admitted, largely to goad people. But for all the glibness, some genetic art plays the provocateur effectively; like the best science fiction, it confronts our assumptions about science. One famous piece consisted solely of one man's sperm DNA in a steel frame, a "portrait" that the artist claimed was "the most realist portrait in [London's National] Portrait Gallery"—because, after all, it revealed the donor's naked DNA. That might seem harshly reductive; but then again, the portrait's "subject" had headed the British arm of arguably the most reductionist biological project ever, the Human Genome Project. Artists have also encoded quotes from Genesis about man's dominion over nature into the A-C-G-T sequence of common bacteria—words that, if the bacteria copy their DNA with high fidelity, could survive mil-lions of years longer than the Bible will. From the ancient Greeks onward, the Pygmalion impulse—the desire to fashion "living" works of art—has driven artists and will only grow stronger as biotechnology advances.

Scientists themselves have even succumbed to the tempta-tion to turn DNA into art. To study how chromosomes wriggle about in three dimensions, scientists have developed ways to "paint" them with fluorescent dyes. And karyotypes—the famil-iar pictures of twenty-three chromosomes paired up like paper dolls—have been transformed from dull dichromatic images into pictures so flamboyantly incandescent that a fauvist would blush. Scientists have also used DNA itself to build bridges, snowflakes, "nano-flasks," kitschy smiley faces, Rock 'Em Sock 'Em Robot look-alikes, and Mercator maps of every continent. There are mobile DNA "walkers" that cartwheel along like a Slinky down the stairs, as well as DNA boxes with lids that open, turned by a DNA "key." Scientist-artists call these fanciful constructions "DNA origami."

To create a piece of DNA origami, practitioners might start with a virtual block on a computer screen. But instead of the block being solid, like marble, it consists of tubes stacked together, like a rectangular bundle of drinking straws. To "carve" something—say, a bust of Beethoven—they first digitally chisel at the surface, removing small segments of tubes, until the leftover tubes and tube fragments have the right shape. Next they thread one long strand of *single*-stranded DNA through every tube. (This threading happens virtually, but the computer uses a DNA strand from a real virus.) Eventually the strand weaves back and forth enough to connect every contour of Beethoven's face and hair. At this point, the sciartists digitally dissolve away the tubes to reveal pure, folded-up DNA, the blueprint for the bust.

To actually build the bust, the sciartists inspect the folded DNA strand. Specifically, they look for short sequences that reside far apart on the unfolded, linear DNA string, but lie close together in the folded configuration. Let's say they find the sequences AAAA and CCCC right near each other. The key step comes now, when they build a separate snippet of real DNA, TTTTGGGG, whose first half complements one of those four-letter sequences and whose second half complements the other. They build this complement base by base using commercial equipment and chemicals, and mix it with the long, unwound viral DNA. At some point the TTTT of the snippet bumps into the AAAA of the long strand, and they lock together. Amid the molecular jostling, the snippet's GGGG will eventually meet and lock onto the CCCC, too, "stapling" the long DNA strand together there. If a unique staple exists for every other joint, too, the sculpture basically assembles itself, since each staple will yank faraway parts of the viral DNA into place. In all, it takes a week to design a sculpture and prepare the DNA. Sciartists then mix the staples and viral DNA, incubate things at 140°F for an hour, and cool it to room temperature over another week. The result: a billion microbusts of Ludwig van.

Beyond the fact that you can spin DNA itself into art, the two intersect on more profound levels. The most miserable societies in human history still found time to carve and color and croon, which strongly implies that evolution wired these impulses into our genes. Even animals show artistic urges. If introduced to painting, chimpanzees often skip feedings to keep smearing canvases and sometimes throw tantrums if scientists take their brushes and palettes away. (Cross, sunburst, and circle motifs dominate their work, and chimps favor bold, Miró-esque lines.) Some monkeys also have musical biases as ruthless as any hipster's,* as do birds. And birds and other creatures are far more discriminating connoisseurs of dance than your average *Homo sapiens*, since many species dance to communicate or court lovers.

Still, it's not clear how to fix such impulses in a molecule. Does "artistic DNA" produce musical RNA? Poetic proteins? What's more, humans have developed art qualitatively different from animal art. For monkeys, an eye for strong lines and symmetry probably helps them craft better tools in the wild, nothing more. But humans infuse art with deeper, symbolic meanings. Those elks painted on cave walls aren't just elks, they're *elks we will hunt tomorrow* or *elk gods*. For this reason, many scientists suspect that symbolic art springs from language, since language teaches us to associate abstract symbols (like pictures and words) with real objects. And given that language has genetic roots, perhaps untangling the DNA of language skills can illuminate the origins of art.

Perhaps. As with art, many animals have hardwired proto-language skills, with their warbles and screeches. And studies of human twins show that around half of the variability in our normal, everyday aptitude with syntax, vocabulary, spelling, listening comprehension—pretty much everything—traces back to DNA. (Linguistic disorders show even stronger genetic correlation.) The problem is, attempts to link linguistic skills or deficits to DNA always run into thickets of genes. Dyslexia, for instance, links to at least six genes, each of which contributes unknown amounts. Even more confusing, similar genetic mutations can produce different effects in different people. So scientists find themselves in the same position as Thomas Hunt Morgan in the fruit fly room. They know that genes and regulatory DNA "for" language exist; but how exactly that DNA enhances our eloquence—increasing neuron counts? sheathing brain cells more efficiently? fiddling with neurotransmitter levels?—no one knows.

Given this disarray, it's easy to understand the excitement, even hype, that attended the recent discovery of a purported master gene for language. In 1990 linguists inferred the gene's existence after studying three generations of a London family known only (for privacy) as the KEs. In a simple pattern of single-gene dominance, half the KEs suffer from a strange suite of language malfunctions. They have trouble coordinating their lips, jaws, and tongues, and stumble over most words, becoming especially incomprehensible on the phone. They also struggle when asked to ape a sequence of simple facial expressions, like opening their mouths, sticking out their tongues, and uttering an *uuuuaaaahh* sound. But some scientists argue that the KEs' problems extend beyond motor skills to grammar. Most of them know the plural of *book* is *books*, but seemingly because they've memorized that fact. Give them made-up words like *zoop* or *wug*, and they cannot figure out the plural; they see no connection between *book/books* and *zoop/zoops*, even after years of language therapy. They also fail fill-in-the-blank tests about the past tense, using words like "bringed." The IQs of affected KEs sink pretty low—86 on average, versus 104 for nonaffected KEs. But the language hiccups probably aren't a simple cognitive deficit: a few afflicted KEs have nonverbal IQ scores above average, and they can spot logical fallacies in arguments when tested. Plus, some scientists found that they understand reflexives just fine (e.g., "he washed him" versus "he washed himself"), as well as passive versus active voice and possessives.

It baffled scientists that one gene could cause such disparate symptoms, so in 1996 they set out to find and decode it. They narrowed its locus down to fifty genes on chromosome seven and were tediously working through each one when they caught a break. Another victim turned up, CS, from an unrelated family. The boy presented with the same mental and mandibular problems, and doctors spotted a translocation in his genes: a Philadelphia-like swap between the arms of two chromosomes, which interrupted the *foxp2* gene on chromosome seven.

Like vitamin A, the protein produced by *foxp2* clamps onto other genes and switches them on. Also like vitamin A, *foxp2* has a long reach, interacting with hundreds of genes and steering fetal development in the jaw, gut, lungs, heart, and especially the brain. All mammals have *foxp2*, and despite billions of years of collective evolution, all versions look pretty much the same; humans have accumulated just three amino acid differences compared to mice. (This gene looks strikingly similar in songbirds as well, and is especially active when they're learning new songs.) Intriguingly, humans picked up two of our amino-acid changes after splitting from chimps, and these changes allow *foxp2* to interact with many new genes. Even more intriguingly, when scientists created mutant mice with the human *foxp2*, the mice had different neuron architecture in a brain region that (in us) processes language, and they conversed with fellow mice in lower-pitched, baritone squeaks.

Conversely, in the affected KEs' brains, the regions that help produce language are stunted and have low densities of neurons. Scientists have traced these deficits back to a single A-for-G mutation. This substitution altered just one of *foxp2*'s 715 amino acids, but it's enough to prevent the protein from binding to DNA. Unfortunately, this mutation occurs in a different part of the gene than the human-chimp mutations, so it can't explain much about the evolution and original acquisition of language. And regardless, scientists still face a cause-and-effect tangle with the KEs: did the neurological deficits cause their facial clumsiness, or did their facial clumsiness lead to brain atrophy by discouraging them from practicing language? *Foxp2* can't be the only language gene anyway, since even the most afflicted in the KE clan aren't devoid of language; they're orders of magnitude more eloquent than any simian. (And sometimes they seemed more creative than the scientists testing them. When presented with the puzzler "Every day he walks eight miles. Yesterday he ___ ," instead of answering, "*walked* eight miles," one afflicted KE muttered, "had a rest.") Overall, then, while *foxp2* reveals something about the genetic basis of language and symbolic thought, the gene has proved frustratingly inarticulate so far.

Even the one thing scientists had all agreed on with *foxp2* — its unique form in humans — proved wrong. *Homo sapiens* split from other *Homo* species hundreds of thousands of years ago, but paleogeneticists recently discovered the human version of *foxp2* in Neanderthals. This might mean nothing. But it might mean that Neanderthals also had the fine motor skills for language, or the cognitive wherewithal. Perhaps both: finer motor skills might have allowed them to use language more, and when they used it more, maybe they found they had more to say.

All that's certain is that the *foxp2* discovery makes another debate about Neanderthals, over Neanderthal art, more urgent. In caves occupied by Neanderthals, archaeologists have discovered flutes made from bear femurs, as well as oyster shells stained red and yellow and perforated for stringing on necklaces. But good luck figuring out what these trinkets meant to Neanderthals. Again, perhaps Neanderthals just aped humans and attached no symbolic meaning to their toys. Or perhaps humans, who often colonized Neanderthal sites after Neanderthals died, simply tossed their worn-out flutes and shells in with Neanderthal rubbish, scrambling the chronology. The truth is, no one has any idea how articulate or artsy-fartsy Neanderthals were.

So until scientists catch another break—find another KE family with different DNA flaws, or root out more unexpected genes in Neanderthals—the genetic origins of language and symbolic art will remain murky. In the meantime we'll have to content ourselves with tracing how DNA can augment, or make a mess of, the work of modern artists.

No different than with athletes, tiny bits of DNA can determine whether budding musicians fulfill their talents and ambitions. A few studies have found that one key musical trait, perfect pitch, gets inherited with the same dominant pattern as the KE language deficit, since people with perfect pitch passed it to half their children. Other studies found smaller and subtler genetic contributions for perfect pitch instead, and found that this DNA must act in concert with environmental cues (like music lessons) to bestow the gift. Beyond the ear, physical attributes can enhance or doom a musician as well. Sergei Rachmaninoff's gigantic hands—probably the result of Marfan syndrome, a genetic disorder—could span twelve inches, an octave and a half on the piano, which allowed him to compose and play music that would tear the ligaments of lesser-endowed pianists. On the other mitt, Robert Schumann's career as a concert pianist collapsed because of focal dystonia—a loss of muscle that caused his right middle finger to curl or jerk involuntarily. Many people with this condition have a genetic susceptibility, and Schumann compensated by writing at least one piece that avoided that finger entirely. But he never let up on his grinding practice schedule, and a jerry-built mechanical rack he designed to stretch the finger may have exacerbated his symptoms.

Still, in the long, gloried history of ailing and invalid musicians, no DNA proved a more ambivalent friend and ambiguous foe than the DNA of nineteenth-century musician Niccolò Paganini, the violin virtuoso's violin virtuoso. The opera composer (and noted epicurean) Gioacchino Rossini didn't like acknowledging that he ever wept, but one of the three times he owned up to crying* was when he heard Paganini perform. Rossini bawled then, and he wasn't the only one bewitched by the ungainly Italian. Paganini wore his dark hair long and performed his concerts in black frock coats with black trousers, leaving his pale, sweaty face hovering spectrally onstage. He also cocked his hips at bizarre angles while performing, and sometimes crossed his elbows at impossible angles in a rush of furious bowing. Some connoisseurs found his concerts histrionic, and accused him of fraying his violin strings before shows so they'd snap dramatically midperformance. But no one ever denied his showmanship: Pope Leo XII named him a Knight of the Golden Spur, and royal mints struck coins with his likeness. Many critics hailed him as the greatest violinist ever, and he has proved almost a singular exception to the rule in classical music that only composers gain immortality.

Paganini rarely if ever played the old masters during his concerts, preferring his own compositions, which highlighted his finger-blurring dexterity. (Ever a crowd-pleaser, he also included lowbrow passages where he mimicked donkeys and roosters on his violin.) Since his teenage years, in the 1790s, Paganini had labored over his music; but he also understood human psychology and so encouraged various legends about the supernatural origins of his gifts. Word got around that an angel had appeared at Paganini's birth and pronounced that no man would ever play the violin so sweetly. Six years later, divine favor seemingly resurrected him from a Lazarus-like doom. After he fell into a cataleptic coma, his parents gave him up for dead—they wrapped him in a burial shroud and everything—when, suddenly, something made him twitch beneath the cloth, saving him by a whisker from premature burial. Despite these miracles, people more often attributed Paganini's talents to necromancy, insisting he'd signed a pact with Satan and exchanged his immortal soul for shameless musical talent. (Paganini fanned these rumors by holding concerts in cemeteries at twilight and giving his compositions names like "Devil's Laughter" and "Witches' Dance," as if he had firsthand experience.) Others argued that he'd acquired his skills in dungeons, where he'd supposedly been incarcerated for eight years for stabbing a friend and had nothing better to do than practice violin. More sober types scoffed at these stories of witchcraft and iniquity. They patiently explained that Paganini had hired a crooked surgeon to snip the motion-limiting ligaments in his hands. Simple as that.

However ludicrous, that last explanation hits closest to the mark. Because beyond Paganini's passion, charisma, and capacity for hard work, he did have unusually supple hands. He could unfurl and stretch his fingers impossibly far, his skin seemingly about to rip apart. His finger joints themselves were also freakishly flexible: he could wrench his thumb across the back of his hand to touch his pinky (try this), and he could wriggle his midfinger joints *laterally*, like tiny metronomes. As a result, Paganini could dash off intricate riffs and arpeggios that other violinists didn't dare, hitting many more high and low notes in swift succession—up to a thousand notes per minute, some claim. He could double- or triple-stop (play multiple notes at once) with ease, and he perfected unusual techniques, like left-handed pizzicato, a plucking technique, that took advantage of his plasticity. Normally the right hand (the bow hand) does pizzicato, forcing the violinist to chose between bowing or plucking during each passage. With left-handed pizzicato, Paganini didn't have to choose. His nimble fingers could bow one note and pluck the next, as if two violins were playing at once.

Beyond being flexible, his fingers were deceptively strong, especially the thumbs. Paganini's great rival Karol Lipiński watched him in concert one evening in Padua, then retired to Paganini's room for a late dinner and some chitchat with Paganini and friends. At the table, Lipiński found a disappointingly scanty spread for someone of Paganini's stature, mostly eggs and bread. (Paganini could not even be bothered to eat that and contented himself with fruit.) But after some wine and some jam sessions on the guitar and trumpet, Lipiński found himself staring at Paganini's hands. He even embraced the master's "small bony fingers," turning them over. "How is it possible," Lipiński marveled, "for these thin small fingers to achieve things acquiring extraordinary strength?" Paganini answered, "Oh, my fingers are stronger than you think." At this he picked up a saucer of thick crystal and suspended it over the table, fingers below, thumb on top. Friends gathered around to laugh—they'd seen the trick before. While Lipiński stared, bemused, Paganini flexed his thumb almost imperceptibly and—*crack!*—snapped the saucer into two shards. Not to be outdone, Lipiński grabbed a plate and tried to shatter it with his own thumb, but couldn't come close. Nor could Paganini's friends. "The saucers remained just as they were before," Lipiński recalled, "while Paganini laughed maliciously" at their futility. It seemed almost unfair, this combination of power and agility, and those who knew Paganini best, like his personal physician, Francesco Bennati, explicitly credited his success to his wonderfully tarantular hands.

Of course, as with Einstein's violin training, sorting out cause and effect gets tricky here. Paganini had been a frail child, sickly and prone to coughs and respiratory infections, but he nevertheless began intensive violin lessons at age seven. So perhaps he'd simply loosened up his fingers through practice. However, other symptoms indicate that Paganini had a genetic condition called Ehlers-Danlos syndrome. People with EDS cannot make much collagen, a fiber that gives ligaments and tendons some rigidity and toughens up bone. The benefit of having less collagen is circus flexibility. Like many people with EDS, Paganini could bend all his joints alarmingly far backward (hence his contortions onstage). But collagen does more than

SIG⁣ᵗ PAGANINI.

Widely considered the greatest violinist ever, Niccolò Paganini owed much of his gift to a genetic disorder that made his hands freakishly flexible. Notice the grotesquely splayed thumb. (Courtesy of the Library of Congress)

prevent most of us from touching our toes: a chronic lack can lead to muscle fatigue, weak lungs, irritable bowels, poor eyesight, and translucent, easily damaged skin. Modern studies have shown that musicians have high rates of EDS and other hypermobility syndromes (as do dancers), and while this gives them a big advantage at first, they tend to develop debilitating knee and back pain later, especially if, like Paganini, they stand while performing.

Constant touring wore Paganini down after 1810, and although he'd just entered his thirties, his body began giving out on him. Despite his growing fortune, a landlord in Naples evicted him in 1818, convinced that anyone as skinny and sickly as Paganini must have tuberculosis. He began canceling engagements, unable to perform his art, and by the 1820s he had to sit out whole years of tours to recuperate. Paganini couldn't have known that EDS underlay his general misery; no doctor described the syndrome formally until 1901. But ignorance only heightened his desperation, and he sought out quack apothecaries and doctors. After diagnosing syphilis and tuberculosis and who knows what else, the docs prescribed him harsh, mercury-based purgative pills, which ravaged his already fragile insides. His persistent cough worsened, and eventually his voice died completely, silencing him. He had to wear blue-tinted shades to shield his sore retinas, and at one point his left testicle swelled, he sobbed, to the size of "a little pumpkin." Because of chronic mercury damage to his gums, he had to bind his wobbly teeth with twine to eat.

Sorting out why Paganini finally died, in 1840, is like asking what knocked off the Roman Empire—take your pick. Abusing mercury drugs probably did the most intense damage, but Dr. Bennati, who knew Paganini before his pill-popping days and was the only doctor Paganini never dismissed in a rage for fleecing him, traced the real problem further back. After examining Paganini, Bennati dismissed the diagnoses of tuberculosis and syphilis as spurious. He noted instead, "Nearly all [Paganini's] later ailments can be traced to the extreme sensitivity of his skin." Bennati felt that Paganini's papery EDS skin left him vulnerable to chills, sweats, and fevers and aggravated his frail constitution. Bennati also described the membranes of Paganini's throat, lungs, and colon—all areas affected by EDS—as highly susceptible to irritation. We have to be cautious about reading too much into a diagnosis from the 1830s, but Bennati clearly traced Paganini's vulnerability to something inborn. And in the light of modern knowledge, it seems likely Paganini's physical talents and physical tortures had the same genetic source.

Paganini's afterlife was no less doomed. On his deathbed in Nice, he refused communion and confession, believing they would hasten his demise. He died anyway, and because he'd skipped the sacraments, during Eastertide no less, the Catholic Church refused him proper burial. (As a result his family had to schlep his body around ignominiously for months. It first lay for sixty days in a friend's bed, before health officials stepped in. His corpse was next transferred to an abandoned leper's hospital, where a crooked caretaker charged tourists money to gawk at it, then to a cement tub in an olive oil processing plant. Family finally smuggled his bones back into Genoa in secret and interred him in a private garden, where he lay for thirty-six years, until the church finally forgave him and permitted burial.*)

Paganini's ex post facto excommunication fueled speculation that church elders had it in for Paganini. He did cut the church out of his ample will, and the Faustian stories of selling his soul couldn't have helped. But the church had plenty of nonfictional reasons to spurn the violinist. Paganini gambled flagrantly, even betting his violin once before a show. (He lost.) Worse, he caroused with maidens, charwomen, and blue-blooded dames all across Europe, betraying a truly capacious appetite for fornication. In his most ballsy conquests, he allegedly seduced two of

Napoleon's sisters, then discarded them. "I am ugly as sin, yet all I have to do," he once bragged, "is play my violin and women fall at my feet." The church was not impressed.

Nevertheless Paganini's hypersexual activity brings up a salient point about genetics and fine arts. Given their ubiquity, DNA probably encodes some sort of artistic impulses—but why? Why should we respond so strongly to the arts? One theory is that our brains crave social interaction and affirmation, and shared stories, songs, and images help people bond. Art, in this view, fosters societal cohesion. Then again, our cravings for art could be an accident. Our brain circuits evolved to favor certain sights, sounds, and emotions in our ancient environment, and the fine arts might simply exploit those circuits and deliver sights, sounds, and emotions in concentrated doses. In this view, art and music manipulate our brains in roughly the same way that chocolate manipulates our tongues.

Many scientists, though, explain our lust for art through a process called sexual selection, a cousin of natural selection. In sexual selection, the creatures that mate the most often and pass on their DNA don't necessarily do so because they have survival advantages; they're simply prettier, sexier. Sexy in most creatures means brawny, well-proportioned, or lavishly decorated—think bucks' antlers and peacocks' tails. But singing or dancing can also draw attention to someone's robust physical health. And painting and witty poetry highlight someone's mental prowess and agility—talents crucial for navigating the alliances and hierarchies of primate society. Art, in other words, betrays a sexy mental fitness.

Now, if talents on par with Matisse or Mozart seem a trifle elaborate for getting laid, you're right; but immodest overabundance is a trademark of sexual selection. Imagine how peacock tails evolved. Shimmering feathers made some peacocks more attractive long ago. But big, bright tails soon became normal, since genes for those traits spread in the next generations. So only males with even bigger and brighter feathers won attention. But again, as the generations passed, everyone caught up. So winning attention required even more ostentation—until things got out of hand. In the same way, turning out a perfect sonnet or carving a perfect likeness from marble (or DNA) might be we thinking apes' equivalent of four-foot plumage, fourteen-point antlers, and throbbing-red baboon derrieres.*

Of course, while Paganini's talents raised him to the apex of European society, his DNA hardly made him worthy stud material: he was a mental and physical wreck. It just goes to show that people's sexual desires can all too easily get misaligned from the utilitarian urge to pass on good genes. Sexual attraction has its own potency and power, and culture can override our deepest sexual instincts and aversions, making even genetic taboos like incest seem attractive. So attractive, in fact, that in certain circumstances, those very perversions have informed and influenced our greatest art.

With Henri Toulouse-Lautrec, painter and chronicler of the Moulin Rouge, his art and his genetic lineage seem as tightly entwined as the strands of a double helix. Toulouse-Lautrec's family traced its line back to Charlemagne, and the various counts of Toulouse ruled southern France as de facto kings for centuries. Though proud enough to challenge the power of popes—who excommunicated the Toulouse-Lautrecs ten separate times—the lineage also produced pious Raymond IV, who for God's glory led a hundred thousand men during the first Crusade to pillage Constantinople and Jerusalem. By 1864, when Henri was born, the family had lost political power but still ruled vast estates, and their lives had settled into a baronial fugue of endless shooting, fishing, and boozing.

Scheming to keep the family lands intact, the various Toulouse-Lautrecs usually married each other. But these consanguineous marriages gave harmful recessive mutations a chance to crawl out from their caves. Every human alive carries a few malignant mutations, and we survive only because we possess two copies of every gene, allowing the good copy to offset the bum one. (With most genes the body gets along just fine at 50 percent of full manufacturing capacity, or even less. *Foxp2*'s protein is an exception.) The odds of two random people both having a deleterious mutation in the same gene sink pretty low, but relatives with similar DNA can easily pass two copies of a flaw to their children. Henri's parents were first cousins; his grandmothers, sisters.

At six months, Henri weighed just ten pounds, and the soft spots on his head reportedly hadn't closed at age four. His skull seemed swollen, too, and his stubby arms and legs attached at odd angles. Even as a teenager he walked with a cane sometimes, but it didn't prevent him from falling, twice, and fracturing both femurs, neither of which healed soundly. Modern doctors can't agree on the diagnosis, but all agree Toulouse-Lautrec suffered from a recessive genetic disorder that, among other pains, left his bones brittle and stunted his lower limbs. (Though he was usually listed as four eleven, estimates of his adult height ranged as low as four foot six—a man's torso propped top-heavy onto child-size legs.) Nor was he the family's only victim. Toulouse-Lautrec's brother died in infancy, and his runtish cousins, also products of consanguineous marriages, had both bone deformities and seizure disorders.*

And honestly, the Toulouse-Lautrecs escaped unscathed compared to other inbred aristocrats in Europe, like the hapless Hapsburg dynasty in seventeenth-century Spain. Like sovereigns throughout history, the Hapsburgs equated incest with bloodline "purity," and they bedded only other Hapsburgs whose pedigrees they knew intimately. (As the saying goes: with nobility,

familiarity breeds.) The Hapsburgs sat on many thrones throughout Europe, but the Iberian branch seemed especially intent on
cousin lovin'—four of every five Spanish Hapsburgs married
family members. In the most backward Spanish villages at the
time, 20 percent of peasant babies usually died. That number
rose to 30 percent among these Hapsburgs, whose mausoleums
were positively stuffed with miscarriages and stillborns, and
another 20 percent of their children died before age ten. The
unlucky survivors often suffered—as seen in royal portraits—
from the "Hapsburg lip," a malformed and prognathous jaw that
left them looking rather apish.* And the cursed lip grew worse
every generation, culminating in the last Spanish Hapsburg
king, pitiful Charles II.

Charles's mother was his father's niece, and his aunt doubled
as his grandmother. The incest in his past was so determined
and sustained that Charles was slightly *more* inbred than a
brother-sister love child would be. The results were ugly in
every sense. His jaw was so misshapen he could barely chew, his
tongue so bloated he could barely speak. The feebleminded
monarch didn't walk until age eight, and although he died just
shy of forty, he had an honest-to-goodness dotage, full of hallu-
cinations and convulsive episodes. Not learning their lesson,
Hapsburg advisers had imported yet another cousin to marry
Charles and bear him children. Mercifully, Charles often ejaculated prematurely and later fell impotent, so no heir was forthcoming, and the dynasty ceased. Charles and other Hapsburg
kings had employed some of the world's great artists to docu-
ment their reigns, and not even Titian, Rubens, and Velázquez
could mask that notorious lip, nor the general Hapsburg decline
across Europe. Still, in an era of dubious medical records, their
beautiful portraits of ugliness remain an invaluable tool to track
genetic decadence and degeneracy.

Despite his own genetic burden, Toulouse-Lautrec escaped the mental wreckage of the Hapsburgs. His wit even won him popularity among his peers—mindful of his bowed legs and limp, boyhood friends often carried him from spot to spot so he could keep playing. (Later his parents bought him an oversized tricycle.) But the boy's father never forgave his son's handicaps. More than anyone else, the strapping, handsome, bipolar Alphonse Toulouse-Lautrec romanticized his family's past. He often dressed up in chain mail like Ray IV, and once lamented to an archbishop, "Ah, Monseigneur! The days are gone when the Counts of Toulouse could sodomize a monk and hang him afterwards if it pleased them." Alphonse bothered having children only because he wanted hunting companions, and after it became clear that Henri would never tramp through a countryside with a gun, Alphonse wrote the boy out of his will.

Instead of hunting, Toulouse-Lautrec took up another family tradition, art. Various uncles had painted with distinction as amateurs, but Henri's interest coursed deeper. From infancy onward he was always doodling and sketching. At a funeral at age three, unable to sign his name yet, he offered to ink an ox into the guest registry instead. And when laid up with broken legs as a teenager, he began drawing and painting seriously. At age fifteen, he and his mother (also estranged from Count Alphonse) moved to Paris so Toulouse-Lautrec could earn a baccalaureate degree. But when the budding man-child found himself in the continent's art capital, he blew off studying and fell in with a crowd of absinthe-drinking bohemian painters. His parents had encouraged his artistic ambitions before, but now their indulgence soured into disapproval over his new, dissolute life. Other family members were outraged. One reactionary uncle dug out Toulouse-Lautrec's juvenilia left behind at the family estate and held a Savonarola-style bonfire of the vanities.

Painter Henri Toulouse-Lautrec, the offspring of first cousins, had a genetic disorder that stunted his growth and subtly shaped his art. He often sketched or painted from unusual points of view. (Henri Toulouse-Lautrec)

But Toulouse-Lautrec had immersed himself in the Paris art scene, and it was then, in the 1880s, that his DNA began to shape his art. His genetic disorder had left him frankly unattractive, bodily and facially—rotting his teeth, swelling his nose, and causing his lips to flop open and drool. To make himself more appealing to women, he masked his face somewhat with a stylish beard and also, like Paganini, encouraged certain rumors. (He allegedly earned the nickname "Tripod" for his stumpy legs and long, you know.) Still, the funny-looking "dwarf" despaired of ever winning a mistress, so he began cruising for women in the slummy Paris bars and bordellos, sometimes disappearing into them for days. And in all of noble Paris, that's where this aristocrat found his inspiration. He encountered scores of tarts and lowlifes, but despite their low status Toulouse-Lautrec took the time to draw and paint them, and his work, even when shading comic or erotic, lent them dignity. He found something human, even noble, in dilapidated bedrooms and back rooms, and unlike his impressionist predecessors, Toulouse-Lautrec renounced sunsets, ponds, sylvan woods, all outdoor scenes. "Nature has betrayed me," he explained, and he forswore nature in return, preferring to have cocktails at hand and women of ill repute posing in front of him.

His DNA likely influenced the type of art he did as well. With his stubby arms, and with hands he mocked as *grosses pattes* (fat paws), manipulating brushes and painting for long stretches couldn't have been easy. This may have contributed to his decision to devote so much time to posters and prints, less awkward mediums. He also sketched extensively. The Tripod wasn't always extended in the brothels, and during his downtime, Henri whipped up thousands of fresh drawings of women in intimate or contemplative moments. What's more, in both these sketches and his more formal portraits of the Moulin Rouge, he often took unusual points of view—drawing figures from below (a "nostril view"), or cutting their legs out of the frame (he loathed dwelling on others' legs, given his own shortcomings), or raking scenes at upward angles, angles that someone of greater physical but lesser artistic stature might never have perceived. One model once remarked to him, "You are a genius of deformity." He responded, "Of course I am."

Unfortunately, the temptations of the Moulin Rouge—casual sex, late nights, and especially "strangling the parakeet," Toulouse-Lautrec's euphemism for drinking himself stupid—depleted his delicate body in the 1890s. His mother tried to dry him out and had him institutionalized, but the cures never took. (Partly because Toulouse-Lautrec had a custom hollowed-out cane made, to fill with absinthe and drink from surreptitiously.) After relapsing again in 1901, Toulouse-Lautrec had a brain-blowing stroke and died from kidney failure just days later, at thirty-six. Given the painters in his glorious family line, he probably had some genes for artistic talent etched inside him; the counts of Toulouse had also bequeathed him his stunted skeleton, and given their equally notable history of dipsomania, they probably gave him genes that contributed to his alcoholism as well. As with Paganini, if Toulouse-Lautrec's DNA made him an artist in one sense, it undid him at last.

PART IV

The Oracle of DNA

Genetics in the Past, Present, and Future

13

The Past Is Prologue — Sometimes

What Can (and Can't) Genes Teach Us About Historical Heroes?

ll of them are past helping, so it's not clear why we bother. But whether it's Chopin (cystic fibrosis?), Dostoyevsky (epilepsy?), Poe (rabies?), Jane Austen (adult chicken pox?), Vlad the Impaler (porphyria?), or Vincent van Gogh (half the *DSM*), we're incorrigible about trying to diagnose the famous dead. We persist in guessing despite a rather dubious record, in fact. Even fictional characters sometimes receive unwarranted medical advice. Doctors have confidently diagnosed Ebenezer Scrooge with OCD, Sherlock Holmes with autism, and Darth Vader with borderline personality disorder.

A gawking fascination with our heroes certainly explains some of this impulse, and it's inspiring to hear how they overcame grave threats. There's an undercurrent of smugness, too: *we* solved a mystery previous generations couldn't. Above all, as one doctor remarked in the *Journal of the American Medical Association* in 2010, "The most enjoyable aspect of retrospective diagnoses [is that] there is always room for debate and, in the face of no definitive evidence, room for new theories and claims." Those claims often take the form of extrapolations—counterfactual sweeps that use mystery illnesses to explain the origins of masterpieces or wars. Did hemophilia bring down tsarist Russia? Did gout provoke the American Revolution? Did bug bites midwife Charles Darwin's theories? But while our amplified knowledge of genetics makes trawling through ancient evidence all the more tempting, in practice genetics often adds to the medical and moral confusion.

For various reasons—a fascination with the culture, a ready supply of mummies, a host of murky deaths—medical historians have pried especially into ancient Egypt and into pharaohs like Amenhotep IV. Amenhotep has been called Moses, Oedipus, and Jesus Christ rolled into one, and while his religious heresies eventually destroyed his dynasty, they also ensured its immortality, in a roundabout way. In the fourth year of his reign in the mid-1300s BC, Amenhotep changed his name to Akhenaten ("spirit of the sun god Aten"). This was his first step in rejecting the rich polytheism of his forefathers for a starker, more monotheistic worship. Akhenaten soon constructed a new "sun-city" to venerate Aten, and shifted Egypt's normally nocturnal religious services to Aten's prime afternoon hours. Akhenaten also announced the convenient discovery that he was Aten's long-lost son. When hoi polloi began grumbling about these changes, he ordered his praetorian thugs to destroy any pictures of deities besides his supposed father, whether on public monuments or some poor family's crockery. Akhenaten even became a grammar nazi, purging all traces of the plural hieroglyphic *gods* in public discourse.

Akhenaten's seventeen-year reign witnessed equally heretical changes in art. In murals and reliefs from Akhenaten's era, the birds, fish, game, and flowers start to look realistic for the first time. Akhenaten's harem of artists also portrayed his royal family—including Nefertiti, his most favored wife, and Tut-ankhamen, his heir apparent—in shockingly mundane domestic scenes, eating meals or caressing and smooching. Yet despite the care to get most details right, the bodies themselves of the royal family members appear grotesque, even deformed. It's all the more mysterious because servants and other less-exalted humans in these portraits still look, well, human. Pharaohs in the past had had themselves portrayed as North African Adonises, with square shoulders and dancers' physiques. Not Akhenaten; amid the otherwise overwhelming naturalism, he, Tut, Nefertiti, and other blue bloods look downright alien.

Archaeologists describing this royal art sound like carnival barkers. One promises you'll "recoil from this epitome of physical repulsiveness." Another calls Akhenaten a "humanoid praying mantis." The catalog of freakish traits could run for pages: almond-shaped heads, squat torsos, spidery arms, chicken legs (complete with knees bending backward), Hottentot buttocks, Botox lips, concave chests, pendulous potbellies, and so on. In many pictures Akhenaten has breasts, and the only known nude statue of him has an androgynous, Ken-doll crotch. In short, these works are the anti-*David*, the anti-*Venus de Milo*, of art history.

As with the Hapsburg portraits, some Egyptologists see the pictures as evidence of hereditary deformities in the pharaonic line. Other evidence dovetails with this idea, too. Akhenaten's older brother died in childhood of a mysterious ailment, and a few scholars believe Akhenaten was excluded from court ceremonies when young because of physical handicaps. And in his son Tut's tomb, amid the plunder, archaeologists discovered 130 walking canes, many showing signs of wear. Unable to resist, doctors have retroactively diagnosed these pharaohs with all sorts of ailments, like Marfan syndrome and elephantiasis. But however suggestive, each diagnosis suffered from a crippling lack of hard evidence.

The Egyptian pharaoh Akhenaten (*seated left*) had his court artists depict him and his family as bizarre, almost alien figures, leading many modern doctors to retrodiagnose Akhenaten with genetic ailments. (Andreas Praefcke)

Enter genetics. The Egyptian government had long hesi-tated to let geneticists have at their most precious mummies. Boring into tissues or bones inevitably destroys small bits of them, and paleogenetics was pretty iffy at first, plagued by con-tamination and inconclusive results. Only in 2007 did Egypt relent, allowing scientists to withdraw DNA from five genera-tions of mummies, including Tut and Akhenaten. When com-bined with meticulous CT scans of the corpses, this genetic work helped resolve some enigmas about the era's art and politics.

First, the study turned up no major defects in Akhenaten or his family, which hints that the Egyptian royals looked like normal people. That means the portraits of Akhenaten—which sure don't look normal—probably didn't strive for verisimilitude. They were propaganda. Akhenaten apparently decided that his status as the sun god's immortal son lifted him so far above the normal human rabble that he had to inhabit a new type of body in public portraiture. Some of Akhenaten's strange features in the pictures (distended bellies, porcine haunches) call to mind fertility deities, so perhaps he wanted to portray himself as the womb of Egypt's well-being as well.

All that said, the mummies did show subtler deformities, like clubbed feet and cleft palates. And each succeeding generation had more to endure. Tut, of the fourth generation, inherited both clubfoot and a cleft palate. He also broke his femur when young, like Toulouse-Lautrec, and bones in his foot died because of poor congenital blood supply. Scientists realized why Tut suffered so when they examined his genes. Certain DNA "stutters" (repetitive stretches of bases) get passed intact from parent to child, so they offer a way to trace lineages. Unfortunately for Tut, both his parents had the same stutters—because his mom and dad had the same parents. Nefertiti may have been Akhenaten's most celebrated wife, but for the crucial business of producing an heir, Akhenaten turned to a sister.

This incest likely compromised Tut's immune system and did the dynasty in. Akhenaten had, one historian noted, "a pathological lack of interest" in anything beyond Egypt, and Egypt's foreign enemies gleefully raided the kingdom's outer edges, imperiling state security. The problem lingered after Akhenaten died, and a few years after the nine-year-old Tut assumed the throne, the boy renounced his father's heresies and restored the ancient gods, hoping for better fortune. It didn't come. While working on Tut's mummy, scientists found scads of malarial DNA deep inside his bones. Malaria wasn't uncommon then; similar tests reveal that both of Tut's grandparents had it, at least twice, and they both lived until their fifties. However, Tut's malarial infection, the scientists argued, "added one strain too many to a body that"—because of incestuous genes—"could no longer carry the load." He succumbed at age nineteen. Indeed, some strange brown splotches on the walls inside Tut's tomb provide clues about just how sudden his decline was. DNA and chemical analysis has revealed these splotches as biological in origin: Tut's death came so quickly that the decorative paint on the tomb's inner walls hadn't dried, and it attracted mold after his retinue sealed him up. Worst of all, Tut compounded his genetic defects for the next generation by taking a half sister as his own wife. Their only known children died at five months and seven months and ended up as sorry swaddled mummies in Tut's tomb, macabre additions to his gold mask and walking sticks.

Powerful forces in Egypt never forgot the family's sins, and when Tut died heirless, an army general seized the throne. He in turn died childless, but another commander, Ramses, took over. Ramses and his successors expunged most traces of Akhenaten, Tut, and Nefertiti in the annals of the pharaohs, erasing them with the same determination Akhenaten had shown in erasing other gods. As a final insult, Ramses and his heirs erected buildings over Tut's tomb to conceal it. In fact, they concealed it so well that even looters struggled to find it. As a result, Tut's treasures survived mostly intact over the centuries—treasures that, in time, would grant him and his heretical, incestuous family something like immortality again.

To be sure, for every well-reasoned retrodiagnosis—Tut, Toulouse-Lautrec, Paganini, Goliath (gigantism, definitely)—there are some doozies. Probably the most egregious retrodiagnosis began in 1962, when a doctor published a paper on porphyria, a group of red-blood-cell disorders.

Porphyria leads to a buildup of toxic by-products that can (depending on the type) disfigure the skin, sprout unwanted body hair, or short-circuit the nerves and induce psychosis. The doctor thought this sounded a lot like werewolves, and he floated the idea that widespread fables about wolf-people might have a medical basis. In 1982 a Canadian biochemist went one better. He noted other symptoms of porphyria — blistering in sunlight, protruding teeth, bloody red urine — and started giving talks suggesting that the disease seemed more likely to inspire tales about vampires. When pressed to explain, he declined to write up a scientific paper, and followed up instead (not a good sign) by appearing on a national U.S. talk show. On Halloween. Viewers heard him explain that "vampire" porphyrics had roamed about at night because of the blistering, and had probably found relief from their symptoms by drinking blood, to replace missing blood components. And what of the famously infectious vampire bite? Porphyric genes run in families, he argued, but it often takes stress or shock to trigger an outbreak. A brother or sister nipping at you and sucking your blood surely qualifies as stressful.

The show drew lots of attention and soon had worried-sick porphyrics asking their physicians if they would mutate into blood-lusting vampires. (A few years later, a deranged Virginia man even stabbed and dismembered his porphyric buddy to protect himself.) These incidents were all the more unfortunate because the theory is hogwash. If nothing else, the traits we consider classically vampiric, like a nocturnal nature, weren't common in folklore vampires. (Most of what we know today are late-nineteenth-century tropes that Bram Stoker invented.) Nor do the supposed scientific facts tally. Drinking blood would bring no relief, since the blood components that cure porphyria don't survive digestion. And genetically, while many porphyrics do suffer from sunburns, the really horrific, blistering burns that might evoke thoughts of supernatural evil are limited to one rare type of porphyria mutation. Just a few hundred cases of it have been documented, ever, far too few to explain the widespread vampire hysteria of centuries past. (Some villages in eastern Europe were tilling their graveyards once per week to search for vampires.) Overall, then, the porphyria fiasco explains more about modern credulity—how willingly people believe things with a scientific veneer—than the origins of folklore monsters.

A more plausible (but still hotly disputed) case of history pivoting around porphyria took place during the reign of Great Britain's King George III. George didn't burn in sunlight but did urinate what looked like rosé wine, among other signs of porphyria, such as constipation and yellowed eyes. He also had rambling fits of insanity. He once solemnly shook hands with an oak branch, convinced he'd finally had the pleasure with the King of Prussia; and in an admittedly vampiric twist, he complained he couldn't see himself in mirrors. At his shrieking worst, ministers confined George to a straitjacket. True, George's symptoms didn't fit porphyria perfectly, and his mental fits were unusually intense for porphyrics. But his genes might have carried complicating factors: hereditary madness was endemic among European royalty between about 1500 and 1900, and most were George's relatives. Regardless of the cause, George suffered his first fit in early 1765, which frightened Parliament enough to pass an act clarifying who should assume power if the king went stark mad. The king, offended, sacked the prime minister. But amid the chaos that spring, the Stamp Act had passed, which began to poison the American colonies' relationship with George. And after a new prime minister took over, the scorned former prime minister decided to focus his remaining power on punishing the colonies, a favorite hobby. Another influential statesman, William Pitt, who wanted to keep America in the empire, might conceivably have blunted that vengeance. But Pitt suffered from another highly heritable disease, gout (possibly triggered by a rich diet or by drinking cheap, lead-tainted Portuguese wines). Laid up, Pitt missed some crucial policy debates in 1765 and afterward, and mad King George's government eventually pushed the American colonists too far.

The new United States rid itself of dynastic lines and bypassed the hereditary insanity that demented European rulers. Of course, U.S. presidents have had their own share of ailments. John F. Kennedy was congenitally sickly—he missed two-thirds of kindergarten owing to illness—and was (incorrectly) diagnosed with hepatitis and leukemia in prep school. When he reached adulthood, doctors sliced open his thigh every two months to insert hormonal pellets, and the family reportedly kept emergency medicine in safety-deposit boxes around the country. Good thing. Kennedy collapsed regularly and received last rites multiple times before becoming president. Historians now know Kennedy had Addison's disease, which ruins the adrenal glands and depletes the body of cortisol. One common side effect of Addison's, bronze skin, might well have supplied Kennedy with his vivacious and telegenic tan.

But it was a serious disease overall, and although his rivals for the presidency in 1960—first Lyndon Johnson, then Richard Nixon—didn't know exactly what ailed JFK, they didn't shy away from spreading rumors that he would (cringe) die during his first term. In response, Kennedy's handlers misled the public through cleverly worded statements. Doctors discovered Addison's in the 1800s as a side effect of tuberculosis; this became known as "classic" Addison's. So Kennedy's people could say with a straight face that he "does not now nor has he ever had an ailment classically described as Addison's disease, which is a tuberculose destruction of the adrenal gland." In truth most cases of Addison's are innate; they're autoimmune attacks coordinated by MHC genes. Moreover, Kennedy probably had at least a genetic susceptibility to Addison's, since his sister Eunice also suffered from it. But short of disinterring Kennedy, the exact genetic contribution (if any) will remain obscure.

With Abraham Lincoln's genetics, doctors have an even trickier case, since they don't know for certain if he suffered from a disease. The first hint that he might have came in 1959 when a physician diagnosed a seven-year-old with Marfan syndrome. After tracking the disease through the boy's family tree, the doctor discovered, eight generations back, one Mordecai Lincoln Jr., the great-great-grandfather of Abe. Although this was suggestive—Lincoln's gaunt physique and spidery limbs look classically Marfan, and it's a dominant genetic mutation, so it runs in families—the discovery proved nothing, since the boy might have inherited the Marfan mutation from any of his ancestors.

The mutated Marfan gene creates a defective version of fibrillin, a protein that provides structural support for soft tissues. Fibrillin helps form the eyes, for instance, so Marfan victims often have poor eyesight. (This explains why some modern doctors diagnosed Akhenaten as Marfanoid; he might naturally have preferred his kingdom's sun god to Egypt's squinting nocturnal deities.) More important, fibrillin girds blood vessels: Marfan victims often die young after their aortas wear out and rupture. In fact, examining blood vessels and other soft tissue was the only sure way to diagnose Marfan for a century. So without Lincoln's soft tissues, doctors in 1959 and after could only pore over photos and medical records and argue about ambiguous secondary symptoms.

The idea of testing Lincoln's DNA emerged around 1990. Lincoln's violent death had produced plenty of skull bits and bloody pillowcases and shirt cuffs to extract DNA from. Even the pistol ball recovered from Lincoln's skull might have traces of DNA. So in 1991 nine experts convened to debate the feasibility, and ethics, of running such tests. Right away a congressman from Illinois (natch) jumped into the fray and demanded that the experts determine, among other things, whether Lincoln would have endorsed the project. This proved difficult. Not only did Lincoln die before Friedrich Miescher even discovered DNA, but Lincoln left no statement (why would he?) of his views on privacy in posthumous medical research. What's more, genetic testing requires pulping small bits of priceless artifacts—and scientists still might not get a solid answer. As a matter of fact, the Lincoln committee realized disturbingly late in the process how complicated getting a diagnosis would be. Emerging work showed that Marfan syndrome could arise from many different fibrillin mutations, so geneticists would have to search through long swaths of DNA to diagnose it—a much harder prospect than searching for a single-point mutation. And if they found nothing, Lincoln still might have Marfan's, through an unknown mutation. Plus, other diseases can mimic Marfan's by garbling other genes, adding to the complications. A serious scientific venture suddenly looked shaky, and it didn't bolster anyone's confidence when garish rumors emerged that a Nobel laureate wanted to clone and hawk "authentic Lincoln DNA" embedded in amber jewelry. The committee eventually scrapped the whole idea, and it remains on hold today.

Though ultimately futile, the attempt to study Lincoln's DNA did provide some guidelines for judging the worth of other retrogenetics projects. The most important scientific consideration is the quality of current technology and whether (despite the frustration of waiting) scientists should hold off and let future generations do the work. Moreover, while it seems obvious that scientists need to prove they can reliably diagnose a genetic disease in living people first, in Lincoln's case they started rushing forward without this assurance. Nor could the technology in 1991 have skirted the inevitable DNA contamination of well-handled artifacts like bloody cuffs and pillowcases. (For this reason, one expert suggested practicing first on the anonymous bones of Civil War amputees piled up in national museums.)

As far as ethical concerns, some scientists argued that historians already invade people's diaries and medical records, and retrogenetics simply extends this license. But the analogy doesn't quite hold, because genetics can reveal flaws even the person in question didn't know about. That's not so terrible if they're safely dead, but any living descendants might not appreciate being outed. And if invading someone's privacy is unavoidable, the work should at least attempt to answer weighty or otherwise unanswerable questions. Geneticists could easily run tests to determine if Lincoln had wet or dry earwax, but that doesn't exactly illuminate Lincoln the man. A diagnosis of Marfan syndrome arguably would. Most Marfan victims die young from a ruptured aorta; so perhaps Lincoln, fifty-six when assassinated,* was doomed to never finish his second term anyway. Or if tests ruled out Marfan, they might point toward something else. Lincoln deteriorated visibly during his last months in office; the *Chicago Tribune* ran an editorial in March 1865 urging him, war or no war, to take time off and rest before stress and overwork killed him. But perhaps it wasn't stress. He might have been afflicted with another Marfan-like disease. And because some of these diseases cause significant pain and even cancer, Lincoln might conceivably have *known* he'd die in office (as FDR later did). This would put Lincoln's change of vice presidents in 1864 and his plans for postwar leniency toward the Confederacy in new light. Genetic tests could also reveal whether the brooding Lincoln had a genetic predilection for depression, a popular but largely circumstantial theory nowadays.

Similar questions apply to other presidents. Given Kennedy's Addison's, perhaps Camelot would have expired prematurely no matter what. (Conversely, Kennedy might not have pushed himself and risen so quickly in politics if he hadn't sensed the

Reaper.) And the genetics of Thomas Jefferson's family brings up fascinating contradictions about his views on slavery.

In 1802 several scurrilous newspapers began suggesting that Jefferson had fathered children with a slave "concubine." Sally Hemings had caught Jefferson's eye in Paris when she served him during his stint as American minister there. (She was probably his late wife's half sister; Jefferson's father-in-law had a slave mistress.) Sometime after returning home to Monticello, Jefferson allegedly took Sally as his lover. Jefferson's newspaper enemies jeered her as the "African Venus," and the Massachusetts legislature publicly debated Jefferson's morals, including the Hemings affair, in 1805. But even friendly eyewitnesses recall that Sally's sons in particular were cocoa-colored doppelgängers of Jefferson. A guest once caught one Hemings boy in tableau behind Jefferson's shoulder during a dinner party, and the resemblance flabbergasted him. Through diaries and other documents, historians later determined that Jefferson was in residence at Monticello nine months before the birth of each of Sally's children. And Jefferson emancipated each of those children at age twenty-one, a privilege he did not extend to other slaves. After moving away from Virginia, one of those freed slaves, Madison, bragged to newspapers that he knew Jefferson was his father, and another, Eston, changed his surname to Jefferson partly because of his resemblance to statues of TJ in Washington, D.C.

Jefferson always denied fathering any slave children, however, and many contemporaries didn't believe the charges either; some blamed nearby cousins instead, or other Jefferson relatives. So in the late 1990s scientists effectively hooked Jefferson up to a genetic polygraph. Because the Y chromosome cannot cross over and recombine with other chromosomes, males pass down the full, unchanged Y to each son. Jefferson had no recognized sons, but other male relatives with the same Jefferson Y, like his uncle, Field Jefferson, did have sons. Field Jefferson's sons had their own sons in turn, and they their own sons, and the Jefferson Y eventually got passed down to a few males living today. Luckily, Eston Hemings's line had also produced males every generation, and geneticists tracked members of both families down in 1999. Their Ys were perfect matches. Of course, the test proved only that *a* Jefferson had fathered Sally Hemings's children, not which specific Jefferson. But given the additional historical evidence, a case against Jefferson for child support looks strong.

Again, it's undeniably titillating to speculate about Jefferson's private life—love blossoming in Paris, his pining for Sally while stuck in sultry D.C.—but *l'affaire* also illuminates Jefferson's character. He would have fathered Eston Hemings in 1808, six years after the first allegations appeared—which reveals either enormous hubris or sincere devotion to Sally. And yet like many of the English monarchs he despised, Jefferson disavowed his bastard children, to salvage his reputation. Even more uncomfortably, Jefferson publicly opposed, and authored legislation to make illegal, marriages between blacks and whites, pandering to fears of miscegenation and racial impurity. It seems a damning account of hypocrisy in perhaps our most philosophical president.

Since the Jefferson revelations, Y-chromosome testing has become an increasingly crucial tool in historical genetics. This does have one downside, in that the patrilineal Y defines someone rather narrowly: you can learn about only one of his many, many ancestors in any generation. (Similar limitations arise with matrilineal mtDNA.) Despite this caveat Y can reveal a surprising amount. For example, Y testing reveals that the biggest biological stud in history probably wasn't Casanova or King Solomon but Genghis Khan, the ancestor of sixteen million men today: one in two hundred males on earth carries his testes-determining chromosome. When the Mongols conquered a territory, they fathered as many children as possible with local women, to tie them to their new overlords. ("It's pretty clear what they were doing when they weren't fighting," one historian commented.) Genghis apparently took on much of this burden himself, and central Asia is littered with his litters today.

Archaeologists have studied the Y and other chromosomes to untangle Jewish history as well. The Old Testament chronicles how Jews once divided themselves into the kingdoms of Judea and Israel, independent states that probably developed distinct genetic markers, since people tended to marry within their extended family. After many millennia of Jewish exiles and diasporas, many historians had given up hope of tracing exactly where the remnants of each kingdom had ended up. But the prevalence of unique genetic signatures (including diseases) among modern Ashkenazi Jews, and other unique genetic signatures among Sephardic and Oriental Jews, has allowed geneticists to trace ancient lineages, and determine that the original biblical divisions have largely persisted through time. Scholars have also traced the genetic origins of Jewish priestly castes. In Judaism, the Cohanim, who all supposedly descended from Aaron, brother of Moses, have special ceremonial roles in temple rites. This honor passes from Cohanim father to Cohanim son, exactly like the Y. And it turns out that Cohanim across the world do indeed have very similar Ys, indicating a single patriarchal line. Further study shows that this "Y-chromosomal Aaron" lived, roughly, in Moses's time, confirming the truth of Jewish tradition. (At least in this case. Levites, a related but distinct Jewish group, also pass down religious privileges in a patrilineal way. But Levites around the world rarely share the same Y, so either Jewish tradition bungled this story,* or Levite wives slept around behind their husbands' backs.)

What's more, studying Jewish DNA has helped confirm a once scarcely credible legend among the Lemba tribesmen in Africa. The Lemba had always maintained that they had Jewish roots—that ages and ages ago, a man named Buba led them out of Israel to southern Africa, where they continue today to spurn pork, circumcise boys, wear yarmulke-like hats, and decorate their homes with emblems of elephants surrounded by six-sided Stars of David. The Buba tale sounded awfully tall to archaeologists, who explained these "Black Hebrews" as a case of cultural transmission, not human emigration. But Lemba DNA ratifies their Jewish roots: 10 percent of Lemba men overall, and half the males in the oldest, most revered families—the priestly caste—have none other than the signature Cohanim Y.

While studying DNA can be helpful in answering some questions, you can't always tell if a famous someone suffered from a genetic disorder just by testing his or her descendants. That's because even if scientists find a clean genetic signal for a syndrome, there's no guarantee the descendants acquired the defective DNA from their celebrated great-great-whatever. That fact, along with the reluctance of most caretakers to disinter ancient bones for testing, leaves many medical historians doing old-fashioned genetic analysis—charting diseases in family trees and piecing together diagnoses from a constellation of symptoms. Perhaps the most intriguing and vexing patient undergoing analysis today is Charles Darwin, because of both the elusive nature of his illness and the possibility that he passed it to his children by marrying a close relative—a potentially heartbreaking example of natural selection in action.

After enrolling in medical school in Edinburgh at age sixteen, Darwin dropped out two years later, when surgery lessons began. In his autobiography Darwin tersely recounted the scenes he endured, but he did describe watching an operation on a sick boy, and you can just imagine the thrashing and screaming in those days before anesthesia. The moment both changed and presaged Darwin's life. Changed, because it convinced him to drop out and do something else for a living. Presaged, because the surgery roiled Darwin's stomach, a premonition of the ill health that dogged him ever after.

His health started to fall apart aboard HMS *Beagle*. Darwin had skipped a prevoyage physical in 1831, convinced he would fail it, and once asea he proved an inveterate landlubber, constantly laid low by seasickness. His stomach could handle only raisins for many meals, and he wrote woeful letters seeking advice from his father, a physician. Darwin did prove himself fit during the *Beagle*'s layovers, taking thirty-mile hikes in South America and collecting loads of samples. But after returning to England in 1836 and marrying, he deteriorated into a honest-to-gosh invalid, a wheezing wreck who often disgusted even himself.

It would take the genius of Akhenaten's greatest court caricaturist to capture how cramped and queasy and out-of-sorts Darwin usually felt. He suffered from boils, fainting fits, heart flutters, numb fingers, insomnia, migraines, dizziness, eczema, and "fiery spokes and dark clouds" that hovered before his eyes. The strangest symptom was a ringing in his ears, after which — as thunder follows lightning — he'd always pass horrendous gas. But above all, Darwin barfed. He barfed after breakfast, after lunch, after dinner, brunch, teatime — whenever — and kept going until he was dry heaving. In peak form he vomited twenty times an hour, and once vomited twenty-seven days running. Mental exertion invariably made his stomach worse, and even Darwin, the most intellectually fecund biologist ever, could make no sense of this. "What thought has to do with digesting roast beef," he once sighed, "I cannot say."

The illness upended Darwin's whole existence. For healthier air, he retreated to Down House, sixteen miles from London, and his intestinal distress kept him from visiting other people's homes, for fear of fouling up their privies. He then invented rambling, unconvincing excuses to forbid friends from calling on him in turn: "I suffer from ill-health of a very peculiar kind," he wrote to one, "which prevents me from all mental excitement, which is always followed by spasmodic sickness, and I do not think I could stand conversation with you, which to me would be so full of enjoyment." Not that isolation cured him. Darwin never wrote for more than twenty minutes without stabbing pain somewhere, and he cumulatively missed years of work with various aches. He eventually had a makeshift privy installed behind a half-wall, half-screen in his study, for privacy's sake— and even grew out his famous beard largely to soothe the eczema always scratching at his face.

That said, Darwin's sickness did have its advantages. He never had to lecture or teach, and he could let T. H. Huxley, his bulldog, do the dirty work of sparring with Bishop Wilberforce and other opponents while he lay about the house and refined his work. Uninterrupted months at home also let Darwin keep up his correspondence, through which he gathered invaluable evidence of evolution. He dispatched many an unwary naturalist on some ridiculous errand to, say, count pigeon tail feathers, or search for greyhounds with tan spots near their eyes. These requests seem strangely particular, but they revealed intermediate evolutionary forms, and in sum they reassured Darwin that natural selection took place. In one sense, then, being an invalid might have been as important to *On the Origin of Species* as visiting the Galápagos.

Darwin understandably had a harder time seeing the benefits of migraines and dry heaving, and he spent years searching for relief. He swallowed much of the periodic table in various medicinal forms. He dabbled in opium, sucked lemons, and took "prescriptions" of ale. He tried early electroshock therapy—a battery-charged "galvanization belt" that zapped his abdomen.

The most eccentric cure was the "water cure," administered by a former classmate from medical school. Dr. James Manby Gully had had no serious plans to practice medicine while in school, but the family's coffee plantation in Jamaica went bust after Jamaican slaves gained their freedom in 1834, and Gully had no choice but to see patients full-time. He opened a resort in Malvern in western England in the 1840s, and it quickly became a trendy Victorian spa; Charles Dickens, Alfred, Lord Tennyson, and Florence Nightingale all took cures there. Darwin decamped to Malvern in 1849 with his family and servants.

The water cure basically consisted of keeping patients as moist as possible at all times. After a 5 a.m. cock-a-doodle-doo, servants wrapped Darwin in wet sheets, then doused him with buckets of cold water. This was followed by a group hike that included plenty of hydration breaks at various wells and mineral springs. Back at their cottages, patients ate biscuits and drank more water, and the completion of breakfast opened up the day to Malvern's main activity, bathing. Bathing supposedly drew blood away from the inflamed inner organs and toward the skin, providing relief. Between baths, patients might have a refreshing cold water enema, or strap themselves into a wet abdominal compress called a "Neptune Girdle." Baths often lasted until dinner, which invariably consisted of boiled mutton, fish, and, obviously, some sparkling local H_2O. The long day ended with Darwin crashing asleep into a (dry) bed.

Would that it worked. After four months at this hydrosanitarium, Darwin felt bully, better than at any time since the *Beagle*, able to hike seven miles a day. Back at Down House he continued the cure in a relaxed form, constructing a sweat lodge for use every morning, followed by a polar-bear plunge into a huge cistern (640 gallons) filled with water as cold as 40°F. But as work piled up on Darwin again, the stress got to him, and the

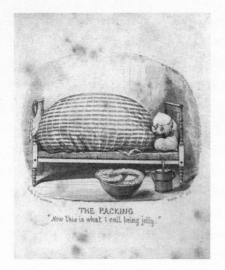

THE PACKING
"Now this is what I call being jolly."

Preparing for the packing.
"Why my nearest and dearest friends
wouldn't know me, I'm a perfect mummy!"

The Douche.
Oh! Oh! Oh! Oh!!!!!!

Scenes from the popular "water cure" in Victorian times, for patients with
stubborn ailments. Charles Darwin underwent a similar regime to cure his
own mystery illness, which dogged him most of his adult life. (Courtesy of
the National Library of Medicine)

water cure lost its power. He relapsed, and despaired of ever knowing the cause of his frailty.

Modern doctors have scarcely done better. A list of more or less probable retrodiagnoses includes middle-ear damage, pigeon allergies, "smoldering hepatitis," lupus, narcolepsy, agoraphobia, chronic fatigue syndrome, and an adrenal-gland tumor. (That last one might explain Darwin's Kennedy-esque bronze glow late in life, despite being a hitherto pasty Englishman who spent most hours indoors.) One reasonably convincing diagnosis is Chagas' disease, which causes flu-like symptoms. Darwin might have picked it up from the South American "kissing bug," since he kept a kissing bug as a pet on the *Beagle*. (He delighted at how it sucked blood from his finger, puffing up like a tick.) But Chagas' disease doesn't fit all Darwin's symptoms. And it's possible that Chagas merely crippled Darwin's digestive tract and left him vulnerable to deeper, dormant genetic flaws. Indeed, other semiplausible diagnoses, like "cyclical vomiting syndrome" and severe lactose intolerance,* have strong genetic components. In addition, much of Darwin's family was sickly growing up, and his mother, Susannah, died of undetermined abdominal trouble when Darwin was eight.

These genetic concerns become all the more poignant because of what happened to Darwin's children. Roughly 10 percent of leisure-class Victorians married blood relatives, and Darwin did his part by marrying Emma Wedgwood, his first cousin. (They shared a grandfather in Josiah Wedgwood, the crockery maven.) Of the ten Darwin children, most were sickly. Three proved infertile as adults, and three died young, roughly double the child mortality rate in England. One, Charles Waring, survived nineteen months; Mary Eleanor lived twenty-three days. When his favorite child, Anne Elizabeth, fell ill, Darwin took her to Dr. Gully for the water cure. When she died anyway, at age ten, it snuffed the last lingering remnant of Darwin's religious faith.

Despite any bitterness toward God, Darwin mostly blamed himself for his children's infirmity. While most children born to first cousins are healthy (well north of 90 percent), they do have higher risks for birth defects and medical problems, and the numbers can creep higher still in unlucky families. Darwin stood uneasily ahead of his time in suspecting this danger. He tested the effects of inbreeding in plants, for instance, not only to shore up his theories of heredity and natural selection but also to see if he could shed light on his own family's ailments. Darwin meanwhile petitioned Parliament to include a question about consanguineous marriages and health in the 1871 census. When the petition failed, the idea continued to fester, and Darwin's surviving children inherited his anxieties. One son, George, argued for outlawing cousin marriages in England, and his (childless) son Leonard presided over the First International Congress of Eugenics in 1912, a congress devoted, ironically enough, to breeding fitter human beings.

Scientists could possibly identify Darwin's ailment with a DNA sample. But unlike Lincoln, Darwin died meekly of a heart attack, leaving no bloody pillowcases. And so far Westminster Abbey refuses to allow DNA sampling of Darwin's bones, partly because doctors and geneticists can't agree what to test for. Complicating things, some doctors conclude that Darwin's illness had a serious hypochondriac edge, too, or sprung from other causes that we can't pin down so easily. Indeed, our focus on Darwin's DNA might even be misplaced, a product of our times. It should serve as a warning that when Freudianism was ascendant, many scientists saw Darwin's illness as the consequence of an Oedipal struggle: the claim was that, unable to overthrow his biological father (an imposing man), Darwin instead "slew the Heavenly Father in the realm of natural history," as one doctor gushed. To such thinking, Darwin's suffering "obviously" derived from repressed guilt for this patricide.

Perhaps our groping about in DNA sequences for the roots of illnesses like Darwin's will look equally quaint someday. And regardless, this groping about misses a deeper point about Darwin and others—that they persevered, despite their illnesses. We tend to treat DNA as a secular soul, our chemical essence. But even a full rendering of someone's DNA reveals only so much.

14

Three Billion Little Pieces

Why Don't Humans Have More Genes Than Other Species?

onsidering its scale, its scope, its ambition, the Human Genome Project—a multidecade, multibillion-dollar effort to sequence all human DNA—was rightly called the Manhattan Project of biology. But few anticipated at the outset that the HGP would be beset with just as many moral ambiguities as the venture in Los Alamos. Ask your biologist friends for a précis of the project, in fact, and you'll get a pretty good handle on their values. Do they admire the project's government scientists as selfless and steadfast or dismiss them as stumbling bureaucrats? Do they praise the private-sector challenge to the government as heroic rebellion or condemn it as greedy self-aggrandizement? Do they think the project succeeded or harp on its disappointments? Like any complex epic, the sequencing of the human genome can support virtually any reading.

The HGP traces its pedigree to the 1970s, when British biologist Frederick Sanger, already a Nobel laureate, invented a method to sequence DNA—to record the order of the A's, C's,

G's, and T's and thereby (hopefully) determine what the DNA does. In brief, Sanger's method involved three basic steps: heating the DNA in question until its two strands separated; breaking those strands into fragments; and using individual A's, C's, G's, and T's to build new complementary strands based on the fragments. Cleverly, though, Sanger sprinkled in special radioactive versions of each base, which got incorporated into the complements. Because Sanger could distinguish whether A, C, G, or T was producing radioactivity at any point along the complement, he could also deduce which base resided there, and tally the sequence.*

Sanger had to read these bases one by one, an excruciatingly tedious process. Nevertheless it allowed him to sequence the first genome, the fifty-four hundred bases and eleven genes of the virus φ-X174. (This work won Sanger a second Nobel in 1980—not bad for someone who once confessed he could never have attended Cambridge University "if my parents had not been fairly rich.") In 1986 two biologists in California automated Sanger's method. And instead of using radioactive bases, they substituted fluorescent versions of A, C, G, and T, each of which produced a different color when strummed by a laser—DNA in Technicolor. This machine, run by a computer, suddenly made large-scale sequencing projects seem feasible.

Strangely, though, the U.S. government agency that funded most biology research, the National Institutes of Health, showed zero interest in DNA sequencing. *Who*, the NIH wondered, *wanted to wade through three billion letters of formless data?* Other departments weren't so dismissive. The Department of Energy considered sequencing a natural extension of its work on how radioactivity damages DNA, and it appreciated the transformative potential of the work. So in April 1987, the DoE opened the world's first human genome project, a seven-year, $1-billion effort centered in Los Alamos, across town from the site of the

Manhattan Project. Funnily enough, as soon as NIH bureaucrats heard the B-word, *billion*, they decided sequencing made
sense after all. So in September 1988 the NIH set up a rival
sequencing institute to scoop up its share of the budgetary pie. In
a scientific coup, it secured James Watson as the institute's chief.

By the 1980s, Watson had developed a reputation as the
"Caligula of biology," someone who, as one science historian put
it, "was given license to say anything that came to his mind and
expect to be taken seriously. And unfortunately he did so, with a
casual and brutal offhandedness." Still, however much he repulsed
some of them personally, Watson retained the intellectual respect
of his colleagues, which proved crucial for his new job, since few
big-name biologists shared his enthusiasm for sequencing. Some
biologists disliked the reductionist approach of the HGP, which
threatened to demote human beings to dribbles of data. Others
feared the project would swallow up all available research funds
but not yield usable results for decades, a classic boondoggle.
Still others simply found the work unbearably monotonous, even
with machines helping. (One scientist cracked that only incarcerated felons should have to sequence—"twenty megabases
[each]," he suggested, "with time off for accuracy.") Most of all,
scientists feared losing autonomy. A project so extensive would
have to be coordinated centrally, and biologists resented the idea
of becoming "indentured servants" who took orders on what
research to pursue. "Many people in the American scientific
community," one early HGP supporter moaned, "will support
small mediocrity before they can even consider the possibility
that there can be some large excellence."

For all his crassness, Watson assuaged his colleagues' fears
and helped the NIH wrest control of the project from the DoE.
He canvassed the country, giving a stump speech about the
urgency of sequencing, and emphasized that the HGP would
sequence not only human DNA but mouse and fruit fly DNA, so

all geneticists would benefit. He also suggested mapping human chromosomes first thing, by locating every gene on them (similar to what Charles Sturtevant did in 1911 with fruit flies). With the map, Watson argued, any scientist could find her pet gene and make progress studying it without waiting fifteen years, the NIH's timeline for sequencing. With this last argument, Watson also had his eye on Congress, whose fickle, know-nothing members might yank funding if they didn't see results last week. To further persuade Congress, some HGP boosters all but promised that as long as Congress ponied up, the HGP would liberate humans from the misery of most diseases. (And not just diseases; some hinted that hunger, poverty, and crime might cease.) Watson brought in scientists from other nations, too, to give sequencing international prestige, and soon the HGP had lumbered to life.

Then Watson, being Watson, stepped in it. In his third year as HGP director, he found out that the NIH planned to patent some genes that one of its neuroscientists had discovered. The idea of patenting genes nauseated most scientists, who argued that patent restrictions would interfere with basic research. To compound the problem, the NIH admitted it had only located the genes it wanted to patent; it had no idea what the genes did. Even scientists who supported DNA patents (like biotech executives) blanched at this revelation. They feared that the NIH was setting a terrible precedent, one that would promote the rapid discovery of genes above everything else. They foresaw a "genome grab," where businesses would sequence and hurriedly patent any gene they found, then charge "tolls" anytime anyone used them for any purpose.

Watson, who claimed that no one had consulted him on all this, went apoplectic, and he had a point: patenting genes could undermine the public-good arguments for the HGP, and it would certainly renew scientists' suspicions. But instead of laying out his concerns calmly and professionally, Watson lit into his boss at the NIH, and behind her back he told reporters the policy was moronic and destructive. A power struggle ensued, and Watson's supervisor proved the better bureaucratic warrior: she raised a stink behind the scenes, Watson alleges, about conflicts of interest in biotech stock he owned, and continued her attempts to muzzle him. "She created conditions by which there was no way I could stay," Watson fumed. He soon resigned.

But not before causing more trouble. The NIH neuroscientist who'd found the genes had discovered them with an automated process that involved computers and robots and little human contribution. Watson didn't approve of the procedure because it could identify only 90 percent of human genes, not the full set. Moreover—always a sucker for elegance—he sneered that the process lacked style and craft. In a hearing before the U.S. Senate about the patents, Watson dismissed the operation as something that "could be run by monkeys." This didn't exactly charm the NIH "monkey" in question, one J. Craig Venter. In fact, partly because of Watson, Venter soon became (in)famous, an international scientific villain. Yet Venter found himself quite suited to the role. And when Watson departed, the door suddenly opened for Venter, perhaps the only scientist alive who was even more polarizing, and who could dredge up even nastier feelings.

Craig Venter started raising hell in childhood, when he'd sneak his bicycle onto airport runways to race planes (there were no fences) and then ditch the cops that chased him. In junior high, near San Francisco, he began boycotting spelling tests, and in high school, his girlfriend's father once held a gun to Venter's head because of the lad's overactive Y chromosome. Later Venter shut down his high school with two days of sit-ins and marches to protest the firing of his favorite teacher—who happened to be giving Venter an F.*

Despite a GPA well below the Mendoza Line, Venter hypnotized himself into believing he would achieve something magnificent in life, but he lacked much purpose beyond that delusion. At twenty-one, in August 1967, Venter joined a M*A*S*H-like hospital in Vietnam as a medic. Over the next year he watched hundreds of men his own age die, sometimes with his hands on them, trying to resuscitate them. The waste of lives disgusted him, and with nothing specific to live for, Venter decided to commit suicide by swimming out into the shimmering-green South China Sea until he drowned. A mile out, sea snakes surfaced around him. A shark also began thumping him with its skull, testing him as prey. As if suddenly waking up, Venter remembered thinking, *What the fuck am I doing?* He turned and scrambled back to shore.

Vietnam stirred in Venter an interest in medical research, and a few years after earning a Ph.D. in physiology in 1975, he landed at the NIH. Among other research, he wanted to identify all the genes our brain cells use, but he despaired over the tedium of finding genes by hand. Salvation came when he heard about a colleague's method of quickly identifying the messenger RNA that cells use to make proteins. Venter realized this information could reveal the underlying gene sequences, because he could reverse-transcribe the RNA into DNA. By automating the technique, he soon cut down the price for detecting each gene from $50,000 to $20, and within a few years he'd discovered a whopping 2,700 new genes.

These were the genes the NIH tried to patent, and the brouhaha established a pattern for Venter's career. He'd get itchy to do something grand, get irritated over slow progress, and find shortcuts. Other scientists would then denounce the work as cheating; one person compared his process for discovering genes to Sir Edmund Hillary taking a helicopter partway up Mount Everest. Whereafter Venter would strongly encourage his detractors to get bent. But his arrogance and gruffness often ended up alienating his allies, too. For these reasons, Venter's reputation grew increasingly ugly in the 1990s: one Nobel laureate jokingly introduced himself by looking Venter up and down and saying, "I thought you were supposed to have horns." Venter had become a sort of Paganini of genetics.

Devil or no, Venter got results. And frustrated by the bureaucracy at the NIH, he quit in 1992 and joined an unusual hybrid organization. It had a nonprofit arm, TIGR (the Institute for Genomic Research), dedicated to pure science. It also had—an ominous sign to scientists—a very-much-for-profit arm backed by a health-care corporation and dedicated to capitalizing on that research by patenting genes. The company made Venter rich by loading him with stock, then loaded TIGR with scientific talent by raiding thirty staff members from the NIH. And true to its rebellious demeanor, once the TIGR team settled in, it spent the next few years refining "whole-genome shotgun sequencing," a radicalized version of Sanger's old-fashioned sequencing methods.

The NIH consortium planned to spend its first few years and its first billion dollars constructing meticulous maps of each chromosome. That completed, scientists would divide each chromosome into segments and send each segment to different labs. Each lab would make copies of the segment and then "shotgun" them—use intense sound waves or another method to blast them into tiny, overlapping bits roughly a thousand bases long. Scientists would next sequence every bit, study how they overlapped, and piece them together into a coherent overall sequence. As observers have noted, the process was analogous to dividing a novel into chapters, then each chapter into sentences. They'd photocopy each sentence and shotgun all the copies into random phrases—"Happy families are all," "are all alike; every unhappy," "every unhappy family is unhappy," and "unhappy in its own way." They would then reconstruct each sentence based on the overlaps. Finally, the chromosome maps, like a book's index, would tell them where their passage was situated overall.

Venter's team loved the shotgun but decided to skip the slow mapping step. Instead of dividing the chromosome into chapters and sentences, they wanted to blast the whole book into overlapping smithereens right away. They'd then whirlwind everything together at once by using banks of computers. The consortium had considered this whole-genome shotgun approach but had dismissed it as slapdash, prone to leaving gaps and putting segments in the wrong place. Venter, however, proclaimed that speed should trump precision in the short term; scientists needed some, any data now, he argued, more than they needed perfect data in fifteen years. And Venter had the fortune to start working in the 1990s, when computer technology exploded and made impatience almost a virtue.

Almost—other scientists weren't so thrilled. A few patient geneticists had been working since the 1980s to sequence the first genome of a fully living creature, a bacterium. (Sanger sequenced only viruses, which aren't fully alive; bacteria have vastly bigger genomes.) These scientists were creeping, tortoise-like, toward finishing their genome, when in 1994 Venter's team began scorching through the two million bases of *Haemophilus influenzae*, another bacterium. Partway through the process, Venter applied for NIH funds to support the work; months later, he received a pink rejection notice, denying him money because of the "impossible" technique he proposed using. Venter laughed; his genome was 90 percent done. And soon afterward the hare won the race: TIGR blew by its poky rivals and published its genome just one year after starting. TIGR completed another full bacterium sequence, of *Mycoplasma genitalium*, just months later. Ever cocky,

Venter not only gloated about finishing both first—and without a red cent from the NIH—he also printed up T-shirts for the second triumph that read I ♥ MY GENITALIUM.

However begrudgingly impressed, HGP scientists had doubts, sensible doubts, that what worked for bacterial DNA would work for the far more complicated human genome. The government consortium wanted to piece together a "composite" genome—a mishmash of multiple men's and women's DNA that would average out their differences and define a Platonic ideal for each chromosome. The consortium felt that only a cautious, sentence-by-sentence approach could sort through all the distracting repeats, palindromes, and inversions in human DNA and achieve that ideal. But microprocessors and sequencers kept getting speedier, and Venter gambled that if his team gathered enough data and let the computers churn, it could beat the consortium. To give due credit, Venter didn't invent shotgunning or write the crucial computer algorithms that pieced sequences together. But he had the hubris (or chutzpah—pick your word) to ignore his distinguished detractors and plunge forward.

And boy did he. In May 1998, Venter announced that he'd cofounded a new company to more or less destroy the international consortium. Specifically, he planned to sequence the human genome in three years—four years before the consortium would finish—and for one-tenth of its $3 billion budget. (Venter's team threw the plans together so quickly the new company had no name; it became Celera.) To get going, Celera's parent corporation would supply it with hundreds of $300,000, state-of-the-science sequencers, machines that (although monkeys could probably run them) gave Venter more sequencing power than the rest of the world combined. Celera would also build the world's largest nonmilitary supercomputer to process data. As a last gibe, even though his work threatened to make them superfluous, Venter suggested to consortium leaders that they could still find valuable work to do. Like sequencing mice.

Venter's challenge demoralized the public consortium. Watson compared Venter to Hitler invading Poland, and most HGP scientists feared they'd fare about as well. Despite their head start, it didn't seem implausible that Venter could catch and pass them. To appease its scientists' demands for independence, the consortium had farmed its sequencing out to multiple U.S. universities and had formed partnerships with labs in Germany, Japan, and Great Britain. With the project so scattered, even some insiders believed the HGP satellites would never finish on time: by 1998, the eighth of the HGP's fifteen years, the groups had collectively sequenced just 4 percent of human DNA. U.S. scientists were especially trembling. Five years earlier, Congress had eighty-sixed the Superconducting Super Collider, a massive particle accelerator in Texas, after delays and overruns had bloated its budget by billions of dollars. The HGP seemed similarly vulnerable.

Key HGP scientists, however, refused to cower. Francis Collins took over the consortium after Watson's resignation, albeit over the objection of some scientists. Collins had done fundamental genetics work at the University of Michigan; he'd found the DNA responsible for cystic fibrosis and Huntington's disease and had consulted on the Lincoln DNA project. He was also fervently Christian, and some regarded him as "ideologically unsound." (After receiving the consortium job offer, Collins spent an afternoon praying in a chapel, seeking Jesus's guidance. Jesus said go for it.) It didn't help matters that, in contrast to the flamboyant Venter, Collins seemed dowdy, once described as having "home-cut hair [and a] Ned Flanders mustache." Collins nevertheless proved politically adept. Right after Venter announced his plans, Collins found himself on a flight with one of Venter's bosses at Celera's money-hungry parent corporation. Thirty thousand feet up, Collins bent the boss's ear, and by the time they landed, Collins had sweet-talked him into supplying the same fancy sequencers to government labs. This pissed Venter off no end. Then, to reassure Congress, Collins announced that the consortium would make the changes necessary to finish the full sequence two years early. It would also release a "rough draft" by 2001. This all sounded grand, but in practical terms, the new timetable forced Collins to eliminate many slower satellite programs, cutting them out of the historic project entirely. (One axed scientist complained of "being treated with K-Y jelly by the NIH" before being you-know-whated guess-where.)

Collins's burly, bearded British counterpart in the consortium was John Sulston, a Cambridge man who'd helped sequence the first animal genome, a worm's. (Sulston was also the sperm donor whose DNA appeared in the supposedly realistic portrait in London.) For most of his career, Sulston had been a lab rat—apolitical, and happiest when holed up indoors and fussing with equipment. But in the mid-1990s, the company that supplied his DNA sequencers began meddling with his experiments, denying Sulston access to raw data files unless he purchased an expensive key, and arguing that it, the company, had the right to analyze Sulston's data, possibly for commercial purposes. In response Sulston hacked the sequencers' software and rewrote their code, cutting the company off. From that moment on, he'd grown wary of business interests and became an absolutist on the need for scientists to exchange DNA data freely. His views became influential when Sulston found himself running one of the consortium's multimillion-dollar labs at the (Fred) Sanger Centre in England. Celera's parent corporation happened to be the same company he'd tangled with before about data, and Sulston viewed Celera itself as Mammon incarnate, certain to hold DNA data hostage and charge researchers exorbitant fees to peruse it. Upon hearing Venter's announcement, Sulston roused his fellow scientists with a veritable St. Crispin's Day speech at a conference. He climaxed by announcing that his institute would double its funding to fight Venter. His troops huzzahed and stomped their feet.

And so it began: Venter versus the consortium. A furious scientific competition, but a peculiar one. Winning was less about insight, reasoning, craft—the traditional criteria of good science—and more about who had the brute horsepower to work faster. Mental stamina was also critical, since the genome competition had, one scientist noted, "all the psychological ingredients of a war." There was an arms race. Each team spent tens of millions to scale up its sequencing power. There was subterfuge. At one point two consortium scientists reviewed for a magazine the fancy new sequencers Celera was using. They gave them a decidedly mixed review—but meanwhile their bosses were secretly negotiating to buy dozens of the machines for themselves. There was intimidation. Some third-party scientists received warnings about their careers being over if they collaborated with Venter, and Venter claims the consortium tried to block publication of his work. There was tension among purported allies. Venter got into innumerable fights with his managers, and a German scientist at one consortium meeting screamed hysterically at Japanese colleagues for making mistakes. There was propaganda. Venter and Celera crowed their every achievement, but whenever they did, Collins would dismiss their "*Mad* magazine" genome, or Sulston would appear on television to argue that Celera had pulled another "con." There was even talk of munitions. After employees received death threats from Luddites, Celera cut down trees near its corporate campus to prevent snipers from nesting in them, and the FBI warned Venter to scan his mail in case a Unabomber wannabe targeted him.

Naturally, the nastiness of the competition titillated the

public and monopolized its attention. But all the while, work of real scientific value was emerging. Under continued criticism, Celera felt it once again had to prove that whole-genome shotgunning worked. So it laid aside its human genome aspirations and in 1999 began sequencing (in collaboration with an NIH-funded team at the University of California, Berkeley) the 120 million bases of the fruit fly genome. To the surprise of many, they produced an absolute beaut: at a meeting just after Celera finished, *Drosophila* scientists gave Venter a standing ovation. And once both teams ramped up their human genome work, the pace was breathtaking. There were still disputes, naturally. When Celera claimed it had surpassed one billion bases, the consortium rejected the claim because Celera (to protect its business interests) didn't release the data for scientists to check. One month later, the consortium itself bragged that it surpassed a billion bases; four months after, it preened over passing two billion. But the harping couldn't diminish the real point: that in just months, scientists had sequenced more DNA, way more, than in the previous two decades combined. Geneticists had excoriated Venter during his NIH days for churning out genetic information without knowing the function. But everyone was playing Venter's game now: blitzkrieg sequencing.

Other valuable insights came when scientists started analyzing all that sequence data, even preliminarily. For one, humans had an awful lot of DNA that looked microbial, a stunning possibility. What's more, we didn't seem to have enough genes. Before the HGP most scientists estimated that, based on the complexity of humans, we had 100,000 genes. In private, Venter remembers a few straying as high as 300,000. But as the consortium and Celera rifled through the genome, that estimate dropped to 90,000, then 70,000, then 50,000—and kept sinking. During the early days of sequencing, 165 scientists had set up a pool with a $1,200 pot for whoever came closest to guessing the correct number of human genes. Usually the entries in a bubble-gum-counting contest like this cluster in a bell curve around the correct answer. Not so with the gene sweepstakes: with every passing day the low guesses looked like the smartest bets.

Thankfully, though, whenever science threatened to become the real HGP story, something juicy happened to distract everyone. For example, in early 2000 President Clinton announced, seemingly out of the clear blue sky, that the human genome belonged to all people worldwide, and he called on all scientists, including ones in the private sector, to share sequence information immediately. There were also whispers of the government eliminating gene patents, and investors with money in sequencing companies stampeded. Celera got trampled, losing $6 billion in stock value—$300 million of it Venter's—in just weeks. As a balm against this and other setbacks, Venter tried around this time to secure a piece of Einstein's brain, to see if someone could sequence its DNA after all,* but the plan came to naught.

Almost touchingly, a few people held out hope that Celera and the consortium could still work together. Sulston had put the kibosh on a cease-fire with Venter in 1999, but shortly thereafter other scientists approached Venter and Collins to broker a truce. They even floated the idea of the consortium and Celera publishing the 90-percent-complete rough draft of the human genome as one joint paper. Negotiations proceeded apace, but government scientists remained chary of Celera's business interests and bristled over its refusal to publish data immediately. Throughout the negotiations, Venter displayed his usual charm; one consortium scientist swore in his face, countless others behind his back. A *New Yorker* profile of Venter from the time opened with a (cowardly anonymous) quote from a senior scientist: "Craig Venter is an asshole." Not surprisingly, plans for a joint publication eventually disintegrated.

Appalled by the bickering, and eyeing an upcoming election,

Bill Clinton finally intervened and convinced Collins and Venter to appear at a press conference at the White House in June 2000. There the two rivals announced that the race to sequence the human genome had ended—in a draw. This truce was arbitrary and, given the lingering resentments, largely bogus. But rather than growling, both Collins and Venter wore genuine smiles that summer day. And why not? It was less than a century after scientists had identified the first human gene, less than fifty years after Watson and Crick had elucidated the double helix. Now, at the millennium, the sequencing of the human genome promised even more. It had even changed the nature of biological science. Nearly three thousand scientists contributed to the two papers that announced the human genome's rough draft. Clinton had famously declared, "The era of big government is over." The era of big biology was beginning.

The two papers outlining the rough draft of the human genome appeared in early 2001, and history should be grateful that the joint publication fell apart. A single paper would have forced the two groups into false consensus, whereas the dueling papers highlighted each side's unique approach—and exposed various canards that had become accepted wisdom.

In its paper, Celera acknowledged that it had poached the free consortium data to help build part of its sequence—which sure undermined Venter's rebel street cred. Furthermore, consortium scientists argued that Celera wouldn't even have finished without the consortium maps to guide the assembly of the randomly shotgunned pieces. (Venter's team published angry rebuttals.) Sulston also challenged the Adam Smith–ish idea that the competition increased efficiency and forced both sides to take innovative risks. Instead, he argued, Celera diverted energy away from sequencing and toward silly public posturing—and sped up the release only of the "fake" rough draft anyway.

Of course, scientists loved the draft, however rough, and the consortium would never have pushed itself to publish one so soon had Venter not flipped his gauntlet at their face. And whereas the consortium had always portrayed itself as the adults here—the ones who didn't care about speedy genomic hot-rodding, just accuracy—most scientists who examined the two drafts side by side proclaimed that Celera did a better job. Some said its sequence was twice as good and less riddled with virus contamination. The consortium also (quietly) put the lie to its criticisms of Venter by copying the whole-genome shotgun approach for later sequencing projects, like the mouse genome.

By then, however, Venter wasn't around to bother the public consortium. After various management tussles, Celera all but sacked Venter in January 2002. (For one thing, Venter had refused to patent most genes that his team discovered; behind the scenes, he was a rather indifferent monomaniacal capitalist.) When Venter left, Celera lost its sequencing momentum, and the consortium claimed victory, loudly, when it alone produced a full human genome sequence in early 2003.*

After years of adrenalized competition, however, Venter, like a fading football star, couldn't simply walk away. In mid-2002 he diverted attention from the consortium's ongoing sequencing efforts by revealing that Celera's composite genome had actually been 60 percent Venter sperm DNA; he had been the primary "anonymous" donor. And, undisturbed by the tsk-tsking that followed his revelation—"vainglorious," "egocentric," and "tacky" were some of the nicer judgments—Venter decided he wanted to analyze his pure DNA, unadulterated by other donors. To this end, he founded a new institute, the Center for the Advancement of Genomics (TCAG, har, har), that would spend $100 million over four years to sequence him and him alone.

This was supposed to be the first complete individual genome—the first genome that, unlike the Platonic HGP genome, included both the mother's and father's genetic contributions, as well as every stray mutation that makes a person unique. But because Venter's group spent four whole years polishing his genome, base by base, a group of rival scientists decided to jump into the game and sequence another individual first—none other than Venter's old nemesis, James Watson. Ironically, the second team—dubbed Project Jim—took a cue from Venter and tried to sweep away the prize with new, cheaper, dirtier sequencing methods, ripping through Watson's full genome in four months and for a staggeringly modest sum, around $2 million. Venter, being Venter, refused to concede defeat, though, and this second genome competition ended, probably inevitably, in another draw: the two teams posted their sequences online within days of each other in summer 2007. The speedy machines of Project Jim wowed the world, but Venter's sequence once again proved more accurate and useful for most research.

(The jockeying for status hasn't ended, either. Venter remains active in research, as he's currently trying to determine [by subtracting DNA from microbes, gene by gene] the minimum genome necessary for life. And however tacky the action might have seemed, publishing his individual genome might have put him in the catbird seat for the Nobel Prize—an honor that, according to the scuttlebutt that scientists indulge in over late-night suds, he covets. A Nobel can be split among three people at most, but Venter, Collins, Sulston, Watson, and others could all make legitimate claims for one. The Swedish Nobel committee would have to overlook Venter's lack of decorum, but if it awards him a solo Nobel for his consistently excellent work, Venter can claim he won the genome war after all.*)

So what did all the HGP competition earn us, science-wise? Depends on whom you ask.

Most human geneticists aim to cure diseases, and they felt certain that the HGP would reveal which genes to target for heart disease, diabetes, and other widespread problems. Congress in fact spent $3 billion largely on this implicit promise. But as Venter and others have pointed out, virtually no genetic-based cures have emerged since 2000; virtually none appear imminent, either. Even Collins has swallowed hard and acknowledged, as diplomatically as possible, that the pace of discoveries has frustrated everyone. It turns out that many common diseases have more than a few mutated genes associated with them, and it's nigh impossible to design a drug that targets more than a few genes. Worse, scientists can't always pick out the significant mutations from the harmless ones. And in some cases, scientists can't find mutations to target at all. Based on inheritance patterns, they know that certain common diseases must have significant genetic components—and yet, when scientists scour the genes of victims of those diseases, they find few if any shared genetic flaws. The "culprit DNA" has gone missing.

There are a few possible reasons for these setbacks. Perhaps the real disease culprits lie in noncoding DNA that lies outside of genes, in regions scientists understand only vaguely. Perhaps the same mutation leads to different diseases in different people because of interactions with their other, different genes. Perhaps the odd fact that some people have duplicate copies of some genes is somehow critically important. Perhaps sequencing, which blasts chromosomes into bits, destroys crucial information about chromosome structure and architectural variation that could tell scientists what genes work together and how. Most scary of all—because it highlights our fundamental ignorance—perhaps the idea of a common, singular "disease" is illusory. When doctors see similar symptoms in different people—fluctuating blood sugar, joint pain, high cholesterol—they naturally assume similar causes. But regulating blood sugar or cholesterol requires scores of genes to work together, and a mutation in any one gene in the cascade could disrupt the whole system. In other words, even if the large-scale symptoms are identical, the underlying genetic causes—what doctors need to pinpoint and treat—might be different. (Some scientists misquote Tolstoy to make this point: perhaps all healthy bodies resemble each other, while each unhealthy body is unhealthy in its own way.) For these reasons, some medical scientists have mumbled that the HGP has— kinda, sorta, so far—flopped. If so, maybe the best "big science" comparison isn't the Manhattan Project but the Apollo space program, which got man to the moon but fizzled afterward.

Then again, whatever the shortcomings (so far) in medicine, sequencing the human genome has had trickle-down effects that have reinvigorated, if not reinvented, virtually every other field of biology. Sequencing DNA led to more precise molecular clocks, and revealed that animals harbor huge stretches of viral DNA. Sequencing helped scientists reconstruct the origins and evolution of hundreds of branches of life, including those of our primate relatives. Sequencing helped trace the global migration of humans and showed how close we came to extinction. Sequencing confirmed how few genes humans have (the lowest guess, 25,947, won the gene sweepstakes), and forced scientists to realize that the exceptional qualities of human beings derive not so much from having special DNA as from regulating and splicing DNA in special ways.

Finally, having a full human genome—and especially having the individual genomes of Watson and Venter—emphasized a point that many scientists had lost sight of in the rush to sequence: the difference between reading a genome and understanding it. Both men risked a lot by publishing their genomes. Scientists across the world pored over them letter by letter, looking for flaws or embarrassing revelations, and each man had different attitudes about this risk. The *apoE* gene enhances our ability to eat meat but also (in some versions) multiplies the risk for Alzheimer's disease. Watson's grandmother succumbed to Alzheimer's years ago, and the prospect of losing his own mind was too much to bear, so he requested that scientists not reveal which *apoE* gene he had. (Unfortunately, the scientists he trusted to conceal these results didn't succeed.*) Venter blocked nothing about his genome and even made private medical records available. This way, scientists could correlate his genes with his height and weight and various aspects of his health—information that, in combination, is much more medically useful than genomic data alone. It turns out that Venter has genes that incline him toward alcoholism, blindness, heart disease, and Alzheimer's, among other ailments. (More strangely, Venter also has long stretches of DNA not normally found in humans but common in chimps. No one knows why, but no doubt some of Venter's enemies have suspicions.) In addition, a comparison between Venter's genome and the Platonic HGP genome revealed far more deviations than anyone expected—four million mutations, inversions, insertions, deletions, and other quirks, any of which might have been fatal. Yet Venter, now approaching seventy years old, has skirted these health problems. Similarly, scientists have noted two places in Watson's genome with two copies of devastating recessive mutations—for Usher syndrome (which leaves victims deaf and blind), and for Cockayne syndrome (which stunts growth and prematurely ages people). Yet Watson, well over eighty, has never shown any hint of these problems.

So what gives? Did Watson's and Venter's genomes lie to us? What's wrong with our reading of them? We have no reason to think Watson and Venter are special, either. A naive perusal of anybody's genome would probably sentence him to sicknesses, deformities, and a quick death. Yet most of us escape. It seems that, however powerful, the A-C-G-T sequence can be circumscribed by extragenetic factors—including our epigenetics.

15

Easy Come, Easy Go?

How Come Identical Twins Aren't Identical?

he prefix *epi-* implies something piggybacking on something else. Epiphyte plants grow on other plants. Epitaphs and epigraphs appear on gravestones and in portentous books. Green things like grass happen to reflect light waves at 550 nm (phenomenon), yet our brains register that light as a *color*, something laden with memory and emotion (epiphenomenon). When the Human Genome Project left scientists knowing almost less than before in some ways—how could twenty-two thousand measly genes, fewer than some grapes have, create complex human beings?— geneticists renewed their emphasis on gene regulation and gene-environment interactions, including *epi*genetics.

Like genetics, epigenetics involves passing along certain biological traits. But unlike genetic changes, epigenetic changes don't alter the hardwired A-C-G-T sequence. Instead epigenetic inheritance affects how cells access, read, and use DNA. (You can think about DNA genes as hardware, epigenetics as software.) And while biology often distinguishes between environment (nurture) and genes (nature), epigenetics combines nature with nurture in novel ways. Epigenetics even hints that we can sometimes inherit the nurture part—that is, inherit biological memories of what our mothers and fathers (or grandmothers and grandfathers) ate and breathed and endured.

Frankly, it's tricky to sort out true epigenetics (or "soft inheritance") from other gene-environment interactions. It doesn't help that epigenetics has traditionally been a grab bag of ideas, the place scientists toss every funny inheritance pattern they discover. On top of everything else, epigenetics has a cursed history, littered with starvation, disease, and suicide. But no other field holds such promise for achieving the ultimate goal of human biology: making the leap from HGP molecular minutiae to understanding the quirks and individuality of full-scale human beings.

Though advanced science, epigenetics actually revives an ancient debate in biology, with combatants who predate Darwin—the Frenchman Jean-Baptiste Lamarck and his countryman, our old friend Baron Cuvier.

Just as Darwin made his name studying obscure species (barnacles), Lamarck cut his teeth on *vermes*. *Vermes* translates to "worms" but in those days included jellyfish, leeches, slugs, octopuses, and other slippery things that naturalists didn't stoop to classify. Lamarck, more discriminating and sensitive than his colleagues, rescued these critters from taxonomic obscurity by highlighting their unique traits and separating them into distinct phyla. He soon invented the term *invertebrates* for this miscellanea, and in 1800 went one better and coined the word *biology* for his whole field of study.

Lamarck became a *biologist* through roundabout means. The moment his pushy father died, Lamarck dropped out of seminary school, bought a creaky horse, and, just seventeen years old, galloped away to join the Seven Years' War. His daughter later claimed that Lamarck distinguished himself there, earning a battlefield promotion to officer, though she often exaggerated his achievements. Regardless, Lt. Lamarck's career ended ignominiously when his men, playing some sort of game that involved lifting Lamarck by his head, injured him. The military's loss was biology's gain, and he soon became a renowned botanist and vermologist.

Not content with dissecting worms, Lamarck devised a grandiloquent theory—the first scientific one—about evolution. The theory had two parts. The overarching bit explained why evolution happened period: all creatures, he argued, had

Jean-Baptiste Lamarck devised perhaps the first scientific theory of evolution. Though mistaken, his theory resembles in some ways the modern science of epigenetics. (Louis-Léopold de Boilly)

"inner urges" to "perfect" themselves by becoming more complex, more like mammals. The second half dealt with the mechanics of evolution, how it occurred. And this was the part that overlaps, conceptually at least, with modern epigenetics, because Lamarck said that creatures changed shape or behavior in response to their environment, then passed those acquired traits on.

For instance, Lamarck suggested that wading shorebirds, straining to keep their derrieres dry, stretched their legs microscopically farther each day and eventually acquired longer legs, which baby birds inherited. Similarly, giraffes reaching to the tippity-tops of trees for leaves acquired long necks and passed those on. It supposedly worked in humans, too: blacksmiths, after swinging hammers year after year, passed their impressive musculature down to their children. Note that Lamarck didn't say that creatures *born* with longer appendages or faster feet or whatever had an advantage; instead creatures worked to develop those traits. And the harder they worked, the better the endowment they passed to their children. (Shades of Weber there, and the Protestant work ethic.) Never a modest man, Lamarck announced the "perfecting" of his theory around 1820.

After two decades of exploring these grand metaphysical notions about life in the abstract, Lamarck's actual physical life began unraveling. His academic position had always been precarious, since his theory of acquired traits had never impressed some colleagues. (One strong, if glib, refutation was that Jewish boys still needed circumcising after three millennia of snip-snipping.) He'd also slowly been going blind, and not long after 1820, he had to retire as professor of "insects, worms, and microscopic animals." Lacking both fame and income, he soon became a pauper, wholly reliant on his daughter's care. When he died in 1829, he could only afford a "rented grave"—meaning his *vermes*-eaten remains got just five years' rest before they were dumped in the Paris catacombs to make room for a new client.

But a bigger posthumous insult awaited Lamarck, courtesy of the baron. Cuvier and Lamarck had actually collaborated when they'd first met in postrevolutionary Paris, if not as friends then as friendly colleagues. Temperamentally, though, Cuvier was 179 degrees opposed to Lamarck. Cuvier wanted facts, facts, facts, and distrusted anything that smacked of speculation— basically all of Lamarck's late work. Cuvier also rejected evolution outright. His patron Napoleon had conquered Egypt and lugged back tons of scientific booty, including murals of animals and mummies of cats, crocodiles, monkeys, and other beasts. Cuvier dismissed evolution because these species clearly hadn't changed in thousands of years, which seemed a good fraction of the earth's lifetime back then.

Rather than limit himself to scientific refutations, Cuvier also used his political power to discredit Lamarck. As one of the many hats he wore, Cuvier composed eulogies for the French scientific academy, and he engineered these *éloges* to ever-so-subtly undermine his deceased colleagues. Poisoning with faint praise, he opened Lamarck's obit by lauding his late colleague's dedication to vermin. Still, honesty compelled Cuvier to point out the many, many times his dear friend Jean-Baptiste had strayed into bootless speculation about evolution. Baron Cuvier also turned Lamarck's undeniable gift for analogies against him, and sprinkled the essay with caricatures of elastic giraffes and damp pelican bums, which became indelibly linked to Lamarck's name. "A system resting on such foundations may amuse the imagination of a poet," Cuvier summed up, "but it cannot for a moment bear the examination of anyone who has dissected the hand, the viscera, or even a feather." Overall, the "eulogy" deserves the title of "cruel masterpiece" that science historian Stephen Jay Gould bestowed. But all morality aside, you do have to hand it to the baron here. To most men, writing eulogies would have been little more than a pain in the neck. Cuvier saw that he could parlay this small burden into a great power, and had the savviness to pull it off.

After Cuvier's takedown, a few romantic scientists clung to Lamarckian visions of environmental plasticity, while others, like Mendel, found Lamarck's theories wanting. Many, though, had trouble making up their minds. Darwin acknowledged in print that Lamarck had proposed a theory of evolution first, calling him a "justly celebrated naturalist." And Darwin did believe that some acquired characteristics (including, rarely, circumcised penises) could be passed down to future generations. At the same time, Darwin dismissed Lamarck's theory in letters to friends as "veritable rubbish" and "extremely poor: I got not fact or idea from it."

One of Darwin's beefs was his belief that creatures gained advantages mostly through inherent traits, traits fixed at birth, not Lamarck's acquired traits. Darwin also emphasized the excruciating pace of evolution, how long everything took, because inborn traits could spread only when the creatures with advantages reproduced. In contrast, Lamarck's creatures took control of their evolution, and long limbs or big muscles spread everywhere lickety-split, in one generation. Perhaps worst, to Darwin and everyone else, Lamarck promoted exactly the sort of empty teleology—mystical notions of animals perfecting and fulfilling themselves through evolution—that biologists wanted to banish from their field forever.*

Equally damning to Lamarck, the generation after Darwin discovered that the body draws a strict line of demarcation between normal cells and sperm and egg cells. So even if a blacksmith had the tris, pecs, and delts of Atlas himself, it doesn't mean squat. Sperm are independent of muscle cells, and if the blacksmith's sperms are 98-milligram weaklings DNA-wise, so too might his children be weaklings. In the 1950s, scientists reinforced this idea of independence by proving that body cells can't alter the DNA in sperm or egg cells, the only DNA that matters for inheritance. Lamarck seemed dead forever.

In the past few decades, though, the *vermes* have turned. Scientists now see inheritance as more fluid, and the barriers between genes and the environment as more porous. It's not all about genes anymore; it's about expressing genes, or turning them on and off. Cells commonly turn DNA off by dotting it with small bumps called methyl groups, or turn DNA on by using acetyl groups to uncoil it from protein spools. And scientists now know that cells pass those precise patterns of methyls and acetyls on to daughter cells whenever they divide—a sort of "cellular memory." (Indeed, scientists once thought that the methyls in neurons physically recorded memories in our brains. That's not right, but interfering with methyls and acetyls can interfere with forming memories.) The key point is that these patterns, while mostly stable, are not permanent: certain environmental experiences can add or subtract methyls and acetyls, changing those patterns. In effect this etches a memory of what the organism was doing or experiencing into its cells—a crucial first step for any Lamarck-like inheritance.

Unfortunately, bad experiences can be etched into cells as easily as good experiences. Intense emotional pain can sometimes flood the mammal brain with neurochemicals that tack methyl groups where they shouldn't be. Mice that are (however contradictory this sounds) bullied by other mice when they're pups often have these funny methyl patterns in their brains. As do baby mice (both foster and biological) raised by neglectful mothers, mothers who refuse to lick and cuddle and nurse. These neglected mice fall apart in stressful situations as adults, and their meltdowns can't be the result of poor genes, since biological and foster children end up equally histrionic. Instead the aberrant methyl patterns were imprinted early on, and as neurons kept dividing and the brain kept growing, these patterns perpetuated themselves. The events of September 11, 2001, might have scarred the brains of unborn humans in similar ways. Some pregnant women in Manhattan developed post-traumatic stress disorder, which can epigenetically activate and deactivate at least a dozen genes, including brain genes. These women, especially the ones affected during the third trimester, ended up having children who felt more anxiety and acute distress than other children when confronted with strange stimuli.

Notice that these DNA changes aren't *genetic*, because the A-C-G-T string remains the same throughout. But epigenetic changes are de facto mutations; genes might as well not function. And just like mutations, epigenetic changes live on in cells and their descendants. Indeed, each of us accumulates more and more unique epigenetic changes as we age. This explains why the personalities and even physiognomies of identical twins, despite identical DNA, grow more distinct each year. It also means that that detective-story trope of one twin committing a murder and both getting away with it—because DNA tests can't tell them apart—might not hold up forever. Their epigenomes could condemn them.

Of course, all this evidence proves only that body cells can record environmental cues and pass them on to other body cells, a limited form of inheritance. Normally when sperm and egg unite, embryos erase this epigenetic information—allowing you to become *you*, unencumbered by what your parents did. But other evidence suggests that some epigenetic changes, through mistakes or subterfuge, sometimes get smuggled along to new generations of pups, cubs, chicks, or children—close enough to bona fide Lamarckism to make Cuvier and Darwin grind their molars.

The first time scientists caught this epigenetic smuggling in action was in Överkalix, a farming hamlet in the armpit between

Sweden and Finland. It was a tough place to grow up during the 1800s. Seventy percent of households there had five or more children—a quarter, ten or more—and all those mouths generally had to be fed from two acres of poor soil, which was all most families could scrape together. It didn't help that the weather above sixty-six degrees north latitude laid waste to their corn and other crops every fifth year or so. During some stretches, like the 1830s, the crops died almost every year. The local pastor recorded these facts in the annals of Överkalix with almost lunatic fortitude. "Nothing exceptional to remark," he once observed, "but that the eighth [consecutive] year of crop failure occurred."

Not every year was wretched, naturally. Sporadically, the land blessed people with an abundance of food, and even families of fifteen could gorge themselves and forget the scarce times. But during those darkest winters, when the corn had withered and the dense Scandinavian forests and frozen Baltic Sea prevented emergency supplies from reaching Överkalix, people slit the throats of hogs and cows and just held on.

This history—fairly typical on the frontier—would probably have gone unremarked except for a few modern Swedish scientists. They got interested in Överkalix because they wanted to sort out whether environmental factors, like a dearth of food, can predispose a pregnant woman's child to long-term health problems. The scientists had reason to think so, based on a separate study of 1,800 children born during and just after a famine in German-occupied Holland—the *Hongerwinter* of 1944–45. Harsh winter weather froze the canals for cargo ships that season, and as the last of many favors to Holland, the Nazis destroyed bridges and roads that could have brought relief via land. The daily ration for Dutch adults fell to five hundred calories by early spring 1945. Some farmers and refugees (including Audrey Hepburn and her family, trapped in Holland during the war) took to gnawing tulip bulbs.

After liberation in May 1945, the ration jumped to two thousand calories, and this jump set up a natural experiment: scientists could compare fetuses who gestated during the famine to fetuses who gestated afterward, and see who was healthier. Predictably, the starved fetuses were generally smaller and frailer babies at birth, but in later years they also had higher rates of schizophrenia, obesity, and diabetes. Because the babies came from the same basic gene pool, the differences probably arose from epigenetic programming: a lack of food altered the chemistry of the womb (the baby's environment) and thereby altered the expression of certain genes. Even sixty years later, the epigenomes of those who'd starved prenatally looked markedly different, and victims of other modern famines—the siege of Leningrad, the Biafra crisis in Nigeria, the Great Leap Forward in Mao's China—showed similar long-term effects.

But because famines had happened so often in Överkalix, the Swedish scientists realized they had an opportunity to study something even more intriguing: whether epigenetic effects could persist through multiple generations. Kings of Sweden had long demanded crop records from every parish (to prevent anyone from cheating on fealties), so agricultural data existed for Överkalix from well before 1800. Scientists could then match the data with the meticulous birth, death, and health records the local Lutheran church kept. As a bonus, Överkalix had very little genetic influx or outflow. The risk of frostbite and a garish local accent kept most Swedes and Lapps from moving there, and of the 320 people the scientists traced, just nine abandoned Överkalix for greener pastures, so scientists could follow families for years and years.

Some of what the Swedish team uncovered—like a link between maternal nutrition and a child's future health—made sense. Much of it didn't. Most notably, they discovered a robust link between a child's future health and a *father's* diet. A father obviously doesn't carry babies to term, so any effect must have slipped in through his sperm. Even more strangely, the child got a health boost only if the father faced starvation. If the father gorged himself, his children lived shorter lives with more diseases.

The influence of the fathers turned out to be so strong that scientists could trace it back to the father's father, too—if grandpa Harald starved, baby grandson Olaf would benefit. These weren't subtle effects, either. If Harald binged, Olaf's risk of diabetes increased fourfold. If Harald tightened his belt, Olaf lived (after adjusting for social disparities) an average of thirty years longer. Remarkably, this was a far greater effect than starvation or gluttony had on Grandpa himself: grandpas who starved, grandpas who gorged, and grandpas who ate just right all lived to the same age, seventy years.

This father/grandfather influence didn't make any genetic sense; famine couldn't have changed the parent's or child's DNA sequence, since that was set at birth. The environment wasn't the culprit, either. The men who starved ended up marrying and reproducing in all different years, so their children and grandchildren grew up in different decades in Överkalix, some good, some bad—yet all benefited, as long as Dad or his dad had done without.

But the influence might make epigenetic sense. Again, food is rich in acetyls and methyls that can flick genes on and off, so bingeing or starving can mask or unmask DNA that regulates metabolism. As for how these epigenetic switches got smuggled between generations, scientists found a clue in the timing of the starvation. Starving during puberty, during infancy, during peak fertility years—none of that mattered for the health of a man's child or grandchild. All that mattered was whether he binged or starved during his "slow growth period," a window from about nine to twelve years old, right before puberty. During this phase, males begin setting aside a stock of cells that will become sperm. So if the slow growth period coincided with a feast or famine, the pre-sperm might be imprinted with unusual methyl or acetyl patterns, patterns that would get imprinted on actual sperm in time.

Scientists are still working out the molecular details of what must have happened at Överkalix. But a handful of other studies about soft paternal inheritance in humans supports the idea that sperm epigenetics has profound and inheritable effects. Men who take up smoking before eleven years old will have tubbier children, especially tubbier boys, than men who start smoking later, even if the grade-school smokers snuff the habit sooner. Similarly, the hundreds of millions of men in Asia and Africa who chew the pulp of betel nuts—a cappuccino-strength stimulant—have children with twice the risk of heart disease and metabolic ailments. And while neuroscientists cannot always find anatomical differences between healthy brains and brains addled with psychoses, they have detected different methyl patterns in the brains of schizophrenics and manic-depressives, as well as in their sperm. These results have forced scientists to revise their assumption that a zygote wipes clean all the environmental tarnish of sperm (and egg) cells. It seems that, Yahweh-like, the biological flaws of the fathers can be visited unto their children, and their children's children.

The primacy of sperm in determining a child's long-term health is probably the most curious aspect of the whole soft inheritance business. Folk wisdom held that maternal impressions, like exposure to one-armed men, was devastating; modern science says paternal impressions count as much or more. Still, these parent-specific effects weren't wholly unexpected, since scientists already knew that maternal and paternal DNA don't quite contribute equally to children. If male lions mount female tigers, they produce a liger—a twelve-foot cat twice as heavy as your average king of the jungle. But if a male tiger knocks up a lion, the resulting tiglon isn't nearly as hefty. (Other mammals show similar discrepancies. Which means that Ilya Ivanov's attempts to impregnate female chimpanzees and female humans weren't as symmetrical as he'd hoped.) Sometimes maternal and paternal DNA even engage in outright combat for control of the fetus. Take the *igf* gene (please).

For once, spelling out a gene's name helps make sense of it: *igf* stands for "insulin-like growth factor," and it makes children in the womb hit their size milestones way earlier than normal. But while fathers want both of a child's *igf* genes blazing away, to produce a big, hale baby that will grow up fast and pass its genes on early and often, mothers want to temper the *igf*s so that baby number one doesn't crush her insides or kill her in labor before she has other children. So, like an elderly couple fighting over the thermostat, sperm tend to snap their *igf* into the on position, while eggs snap theirs off.

Hundreds of other "imprinted" genes turn off or on inside us, too, based on which parent bestowed them. In Craig Venter's genome, 40 percent of his genes displayed maternal/paternal differences. And deleting the exact same stretch of DNA can lead to different diseases, depending on whether Mom's or Dad's chromosome is deficient. Some imprinted genes even switch allegiance over time: in mice (and presumably in humans) maternal genes maintain control over brains as children, while paternal genes take over later in life. In fact, we probably can't survive without proper "epigender" imprinting. Scientists can easily engineer mice embryos with two sets of male chromosomes or two sets of female chromosomes, and according to traditional genetics, this shouldn't be a big deal. But these double-gendered embryos expire in the womb. When scientists mixed in a few cells from the opposite sex to help the embryos survive, the males[2] became huge Botero babies (thanks to *igf*) but had puny brains. Females[2] had small bodies but oversized brains. Variations, then, between the brain sizes of Einstein and Cuvier might be nothing but a quirk of their parents' bloodlines, like male pattern baldness.

So-called parent-of-origin effects have also revived interest in one of the most egregious scientific frauds ever perpetrated. Given the subtlety of epigenetics — scientists have barely gotten a handle in the past twenty years — you can imagine that a scientist stumbling across these patterns long ago would have struggled to interpret his results, much less convince his colleagues of them. And Austrian biologist Paul Kammerer did struggle, in science and love and politics and everything else. But a few epigeneticists today see his story as maybe, just maybe, a poignant reminder about the peril of making a discovery ahead of its time.

Paul Kammerer had an alchemist's ambitions to remake nature, coupled with a teenage boy's talent for harassing small animals. Kammerer claimed he could change the colors of salamanders — or give them polka dots or pinstripes — simply by foisting them into landscapes of unusual hues. He forced sun-loving praying mantises to dine in the dark, and amputated the proboscises of sea squirts just to see the effect on their future children. He even claimed he could grow certain amphibians with or without eyes, depending on how much sunlight they got as youngsters.

Kammerer's triumph, and his undoing, were a series of experiments on the midwife toad, a most peculiar species. Most toads mate in water, then let their fertilized eggs float away freely. Midwife toads make love on land, but because tadpole eggs are more vulnerable on land, the male midwife toad ties the bundle of eggs to his back legs like a bunch of grapes and hops along with them until they hatch. Unmoved by this charming habit, Kammerer decided in 1903 to start forcing midwife toads to breed in water, by cranking the heat way, way up in their aquariums. The tactic worked—the toads would have shriveled like dried apricots if they hadn't spent all their time submerged—and those that survived became more waterlike each generation. They had longer gills, produced a slippery jelly coating to water-proof their eggs, and (remember this) developed "nuptial pads"—black, calluslike growths on their forelimbs, to help male toads grip their slippery mates during aqueous coitus. Most intrigu-ing, when Kammerer returned these abused toads to cooler and moister tanks and let them reproduce, the toads' *descendants* (who never experienced the desert conditions) supposedly inherited the water-breeding preferences and passed them along to still more descendants.

Kammerer announced these results around 1910. Over the next decade, he used this and other experiments (and no experi-ment of his ever failed, it seemed) to argue that animals could be molded to do or be almost anything, given the proper environ-ment. Saying such things at the time had deep Marxist implica-tions, since Marxism held that the only thing keeping the wretched masses down was their terrible environment. But as a committed socialist, Kammerer readily extended his arguments to human society: to his thinking, nurture *was* nature, a unified concept.

Indeed, while biology itself was in serious confusion at the time—Darwinism remained controversial, Lamarckism was all but dead, Mendel's laws hadn't yet triumphed—Kammerer prom-ised he could unite Darwin, Lamarck, and Mendel. For instance, Kammerer preached that the proper environment could actually cause advantageous genes to spring into existence. And far from scoffing, people lapped his theories up; his books became best-sellers, and he lectured to SRO audiences worldwide. (In these "big-show talks" Kammerer also suggested "curing" homosexu-als with testicle transplants and enacting American-style Prohi-bition worldwide, since Prohibition would undoubtedly produce

Paul Kammerer, a tormented Austrian biologist who perpetrated one of the great frauds in science history, may have been an unwitting pioneer in epigenetics. (Courtesy of the Library of Congress)

a generation of American Übermenschen, a race "born without any desire for liquor.")

Unfortunately, the more prominent Kammerer became—he soon anointed himself a "second Darwin"—the shakier his science looked. Most disturbing, Kammerer had withheld crucial details about his amphibian experiments in his scientific reports. Given his ideological posturing, many biologists thought he was blowing smoke, especially William Bateson, Mendel's bulldog in Europe.

A ruthless man, Bateson never shied away from attacking

other scientists. During the eclipse of Darwinism around 1900, he got into an especially nasty row with his former mentor, a Darwin defender named Walter Weldon. Bateson quickly went Oedipal on Weldon by landing himself on the board of a scientific society that allocated biology funding, then cutting Weldon off. Things got so bad later that when Weldon died in 1906, his widow blamed Bateson's rancor for the death, even though Weldon died from a heart attack while cycling. Meanwhile a Weldon ally, Karl Pearson, blocked Bateson's papers from appearing in journals, and also attacked Bateson in his (Pearson's) house organ, a journal called *Biometrika*. When Pearson refused Bateson the courtesy of responding in print, Bateson printed up fake copies of *Biometrika* complete with facsimile covers, inserted his response inside, and distributed them to libraries and universities without any indication they were fraudulent. A limerick at the time summed things up thus: "Karl Pearson is a biometrician / and this, I think, is his position. / Bateson and co. / [I] hope they may go / to monosyllabic perdition."

Now Bateson demanded a chance to examine Kammerer's toads. Kammerer defied him by refusing to supply specimens, and critics continued to roast Kammerer, unimpressed with his excuses. The chaos of World War I temporarily halted the debate, as it left Kammerer's lab in shambles and his animals dead. But as one writer put it, "If World War I did not completely ruin Austria and Kammerer with it, Bateson moved in after the war to finish the job." Under relentless pressure, Kammerer finally, in 1926, let an American ally of Bateson's examine the only midwife toad he'd preserved. This biologist, reptile expert Gladwyn Kingsley Noble, reported in *Nature* that the toad looked entirely normal, except for one thing. The nuptial pads weren't present. However, someone had injected black ink under the toad's skin with a syringe, to make it appear they were. Noble didn't use the word *fraud*, but he didn't have to.

Biology erupted. Kammerer denied any wrongdoing, alluding to sabotage by unnamed political enemies. But the howling of other scientists increased, and Kammerer despaired. Just before the damning *Nature* paper, Kammerer had accepted a post in the Soviet Union, a state favorable to his neo-Lamarckian theories. Six weeks later, Kammerer wrote to Moscow that he couldn't in good conscience accept the job anymore. All the negative attention on him would reflect badly on the great Soviet state.

Then the resignation letter took a dark turn. "I hope I shall gather together enough courage and strength," Kammerer wrote, "to put an end [to] my wrecked life tomorrow." He did, shooting himself in the head on September 23, 1926, on a rocky rural trail outside Vienna. It seemed a sure admission of guilt.

Still, Kammerer always had his defenders, and a few historians have built a not unreasonable case for his innocence. Some experts believe that nuptial pads actually appeared, and that Kammerer (or an overzealous assistant) injected ink merely to "touch up" the evidence. Others believe that political opponents did frame Kammerer. The local national socialist party (precursor of the Nazi party) supposedly wanted to tarnish Kammerer, who was part Jewish, because his theories cast doubt on the inborn genetic superiority of Aryans. What's more, the suicide can't necessarily be pinned on Noble's exposé. Kammerer had chronic money problems and had already become mentally unhinged over one Alma Mahler Gropius Werfel. Werfel worked as Kammerer's unpaid lab assistant for a spell, but she's best known as the tempestuous ex-wife of (among others) composer Gustav Mahler.* She had a fling with the geeky Kammerer, and while it was just another lay to her, Kammerer grew obsessed. He once threatened to blow his brains out all over Mahler's tombstone if she didn't marry him. She laughed.

On the other hand, a prosecutor in any Kammerer case could point out some uncomfortable facts. First, even the unscientific

Werfel, a socialite and dilettante composer of light ditties, recognized that Kammerer was sloppy in the lab, keeping terrible
records and constantly (albeit unconsciously, she felt) ignoring
results that contradicted his pet theories. Even more damning,
scientific journals had caught Kammerer fudging data before.
One scientist called him "the father of photographic image
manipulation."

Regardless of Kammerer's motive, his suicide ended up smearing Lamarckism by association, since nasty political types in the
Soviet Union took up Kammerer's cause. Officials first decided
to shoot an agitprop film to defend his honor. *Salamandra* tells
the story of a Kammerer-like hero (Professor Zange) undone by
the machinations of a reactionary priest (a stand-in for Mendel?).
The priest and an accomplice sneak into Zange's lab and inject
ink into a salamander one night; the next day, Zange is humiliated when someone dunks the specimen into a bath in front of
other scientists and the ink leaks out, clouding the water. After
losing his job, Zange ends up begging for food in the streets
(accompanied, oddly, by a monkey rescued from an evil lab). But
just as he decides to off himself, a woman rescues him and drags
him away to the Soviet paradise. As laughable as this sounds, the
soon-to-be agricultural czar of the Soviet Union, Trofim Lysenko,
basically believed the myth: he considered Kammerer a martyr
to socialist biology and began championing Kammerer's theories.

Or at least parts of them. With Kammerer conveniently
dead, Lysenko could emphasize only his neo-Lamarckian ideas,
which suited Soviet ideology better. And burning with Lamarckian zeal, Lysenko rose to power in the 1930s and began liquidat-
ing scores of non-Lamarckian geneticists (including a protégé of
Bateson), either having them killed outright or starving them in
the Gulag. Unfortunately, the more people he disappeared, the
more Soviet biologists had to pay fealty to Lysenko's twisted
ideas. A British scientist at the time reported that talking to

Lysenko about genetics "was like trying to explain the differential calculus to a man who did not know his twelve times table. He was...a biological circle-squarer." Not surprisingly, Lysenkoism destroyed Soviet agriculture—millions died in famines—but officials refused to abandon what they saw as the spirit of Kammerer.

However unfair, the association with the Kremlin doomed both Kammerer's reputation and Lamarckism in the following decades, though Kammerer's defenders continued to plead his case. Most notably (and ironically, given his denunciation of communism elsewhere) in 1971 novelist Arthur Koestler wrote a nonfiction book, *The Case of the Midwife Toad*, to exonerate Kammerer. Among other things, Koestler dug up a 1924 paper about the discovery of a wild midwife toad *with nuptial pads*. This doesn't necessarily clear Kammerer but does hint that midwife toads have latent genes for nuptial pads. Perhaps a mutation during Kammerer's experiments had brought them out.

Or perhaps epigenetics had. Some scientists have noted recently that, among other effects, Kammerer's experiments changed the thickness of the gelatinous coat that surrounds midwife toad eggs. Because this jelly is rich in methyls, changing the thickness might switch genes on or off, including atavistic genes for nuptial pads or other traits. Equally intriguing, whenever Kammerer mated toads, he insisted that the father's land/water breeding preference "undisputedly" dominated over the female's preference in the next generations. If Dad liked dry sex, ditto his children and grandchildren, and the same proved true if Dad preferred water sex. Such parent-of-origin effects play an important role in soft inheritance, and these toad trends echo those from Överkalix.

To be sure, even if Kammerer did stumble onto epigenetic effects, he didn't understand them—and he probably still (unless you buy the proto-Nazi conspiracy) committed fraud by injecting

ink. But in some ways, that makes Kammerer all the more fascinating. His record of bluster, propaganda, and scandal helps
explain why many scientists, even during the chaotic eclipse of
Darwinism, refused to consider epigenetic-like theories of soft
inheritance. Yet Kammerer might have been both a scoundrel
and an unwitting pioneer: someone willing to lie for a greater
ideological truth—but someone who may not have been lying
after all. Regardless, he grappled with the same issues geneticists
still grapple with today—how the environment and genes interact, and which one, if either, dominates in the end. Indeed, it's
poignant to wonder how Kammerer might have reacted if only
he'd known about, say, Överkalix. He was living and working in
Europe just when some of the transgenerational effects were
emerging in the Swedish village. Fraud or no fraud, had he seen
even traces of his beloved Lamarckism, he might not have felt
desperate enough to take his own life.

Epigenetics has expanded so rapidly in the past decade that trying to catalog every advance can get pretty overwhelming. Epigenetic mechanisms do things as frivolous as give mice polka-dot
tails—or as serious as push people toward suicide (perhaps a
final irony in the Kammerer case). Drugs like cocaine and her-
oin seem to spool and unspool the DNA that regulates neurotransmitters and neurostimulants (which explains why drugs
feel good), but if you keep on chasing the dragon, that DNA can
become permanently misspooled, leading to addiction. Restoring acetyl groups in brain cells has actually resurrected forgotten memories in mice, and more work emerges every day showing
that tumor cells can manipulate methyl groups to shut off the
genetic governors that would normally arrest their growth.
Some scientists think they can even tease out information about
Neanderthal epigenetics someday.

All that said, if you want to make a biologist cranky, start expounding about how epigenetics will rewrite evolution or help us escape our genes, as if they were fetters. Epigenetics does alter how genes function, but doesn't vitiate them. And while epigenetic effects certainly exist in humans, many biologists suspect they're easy come, easy go: methyls and acetyls and other mechanisms might well evaporate within a few generations as environmental triggers change. We simply don't know yet whether epigenetics can *permanently* alter our species. Perhaps the underlying A-C-G-T sequence always reasserts itself, a granite wall that emerges as the methyl-acetyl graffiti wears away.

But really, such pessimism misses the point, and promise, of epigenetics. The low genetic diversity and low gene count of human beings seem unable to explain our complexity and variety. The millions upon millions of different combinations of epigenes just might. And even if soft inheritance evaporates after, say, a half-dozen generations, each one of us lives for two or three generations only—and on those timescales, epigenetics makes a huge difference. It's much easier to rewrite epigenetic software than to rewire genes themselves, and if soft inheritance doesn't lead to true genetic evolution, it does allow us to adapt to a rapidly shifting world. As a matter of fact, thanks to the new knowledge that epigenetics lends us—about cancer, about cloning, about genetic engineering—our world will likely shift even more rapidly in the future.

16

Life as We Do (and Don't) Know It

What the Heck Will Happen Now?

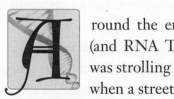round the end of the 1950s, a DNA biochemist (and RNA Tie Club member) named Paul Doty was strolling through New York, minding his own, when a street vendor's wares caught his eye, and he halted, bewildered. The vendor sold lapel buttons, and among the usual crude assortment, Doty noticed one that read "DNA." Few people worldwide knew more about DNA than Doty, but he assumed the public knew little about his work and cared less. Convinced the initialism stood for something else, Doty asked the vendor what D-N-A might be. The vendor looked the great scientist up and down. "Get with it, bud," he barked in New Yawk brogue. "Dat's da gene!"

Jump forward four decades to the summer of 1999. Knowledge of DNA had mushroomed, and Pennsylvania legislators, stewing over the impending DNA revolution, asked a bioethics expert (and Celera board member) named Arthur Caplan to advise them on how lawmakers might regulate genetics. Caplan obliged, but things got off to a rocky start. To gauge his audience, Caplan opened with a question: "Where are your genes?"

Where are they located in the body? Pennsylvania's best and brightest didn't know. With no shame or irony, one quarter equated their genes with their gonads. Another overconfident quarter decided their genes resided in their brains. Others had seen pictures of helixes or something but weren't sure what that meant. By the late 1950s, the term *DNA* was enough a part of the zeitgeist to grace a street vendor's button. *Dat's da gene.* Since then public understanding had plateaued. Caplan later decided, given their ignorance, "Asking politicians to make regulations and rules about genetics is dangerous." Of course, befuddlement or bewilderment about gene and DNA technology doesn't prevent anyone from having strong opinions.

That shouldn't surprise us. Genetics has fascinated people practically since Mendel tilled his first pea plant. But a parasite of revulsion and confusion feeds on that fascination, and the future of genetics will turn on whether we can resolve that push-pull, gotta-have-it-won't-stand-for-it ambivalence. We seem especially mesmerized/horrified by genetic engineering (including cloning) and by attempts to explain rich, complicated human behavior in terms of "mere" genes—two often misunderstood ideas.

Although humans have been genetically engineering animals and plants since the advent of agriculture ten thousand years ago, the first explicit genetic engineering began in the 1960s. Scientists basically started dunking fruit fly eggs in DNA goo, hoping that the porous eggs would absorb something. Amazingly these crude experiments worked; the flies' wings and eyes changed shape and color, and the changes proved heritable. A decade later, by 1974, a molecular biologist had developed tools to splice DNA from different species together, to form hybrids. Although this Pandora restricted himself to microbes, some biologists saw these chimeras and shivered—who knew what was next? They decided that scientists had gotten ahead of themselves, and called for a moratorium on this recombinant

DNA research. Remarkably, the biology community (including the Pandora) agreed, and voluntarily stopped experimenting to debate safety and rules of conduct, almost a unique event in science history. By 1975 biologists decided they did understand enough to proceed after all, but their prudence reassured the public.

That glow didn't last. Also in 1975, a slightly dyslexic myrmecologist born in evangelical Alabama and working at Harvard published a six-pound, 697-page book called *Sociobiology*. Edward O. Wilson had labored for decades in the dirt over his beloved ants, figuring out how to reduce the byzantine social interactions of serfs, soldiers, and queens into simple behavioral laws, even precise equations. In *Sociobiology* the ambitious Wilson extended his theories to other classes, families, and phyla, ascending the evolutionary ladder rung by rung to fish, birds, small mammals, mammalian carnivores, and primates. Wilson then plowed straight through chimps and gorillas to his notorious twenty-seventh chapter, "Man." In it, he suggested that scientists could ground most if not all human behavior—art, ethics, religion, our ugliest aggressions—in DNA. This implied that human beings were not infinitely malleable but had a fixed nature. Wilson's work also implied that some temperamental and social differences (between, say, men and women) might have genetic roots.

Wilson later admitted he'd been politically idiotic not to anticipate the firestorm, maelstrom, hurricane, and plague of locusts that such suggestions would cause among academics. Sure enough, some Harvard colleagues, including the publicly cuddly Stephen Jay Gould, lambasted *Sociobiology* as an attempt to rationalize racism, sexism, poverty, war, a lack of apple pie, and everything else decent people abhor. They also explicitly linked Wilson with vile eugenics campaigns and Nazi pogroms—then acted surprised when other folks lashed out. In 1978, Wilson was defending his work at a scientific conference when a few half-wit activists stormed onstage. Wilson, in a wheelchair with a broken ankle, couldn't dodge or fight back, and they wrested away his microphone. After charging him with "genocide," they poured ice water over his head, and howled, "You're all wet."

By the 1990s, thanks to its dissemination by other scientists (often in softer forms), the idea that human behavior has firm genetic roots hardly seemed shocking. Similarly, we take for granted today another sociobiological tenet, that our hunter-scavenger-gatherer legacy left us with DNA that still biases our thinking. But just as the sociobiology ember was flickering, scientists in Scotland spurted kerosene on the public's fear of genetics by announcing, in February 1997, the birth of probably the most famous nonhuman animal ever. After transferring adult sheep DNA into four hundred sheep eggs, then zapping them *Frankenstein*-style with electricity, the scientists managed to produce twenty viable embryos—clones of the adult donor. These clones spent six days in test tubes, then 145 in utero, during which time nineteen spontaneously aborted. Dolly lived.

In truth, most of the humans gawking at this little lamb cared nothing about Dolly qua Dolly. The Human Genome Project was rumbling along in the background, promising scientists a blueprint of humanity, and Dolly stoked fears that scientists were ramping up to clone one of our own—and with no moratorium in sight. This frankly scared the bejeezus out of most people, although Arthur Caplan did field one excited phone call about the possibility of cloning Jesus himself. (The callers planned to lift DNA from the Shroud of Turin, natch. Caplan remembered thinking, "You are trying to bring back one of the few people that are supposed to come back anyway.")

Dolly's pen mates accepted her, and didn't seem to care about her ontological status as a clone. Nor did her lovers—she eventually gave birth to six (naturally begotten) lambs, all strapping.

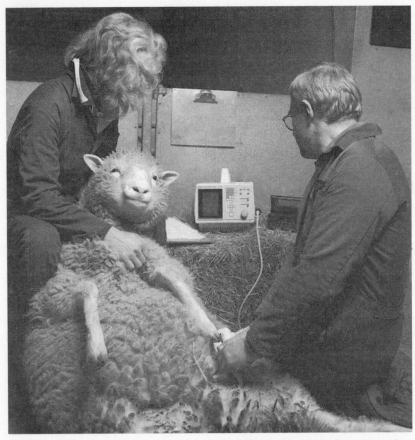

Dolly, the first cloned mammal, undergoes a checkup. (Photo courtesy of the Roslin Institute, University of Edinburgh)

But for whatever reason, human beings fear clones almost instinctively. Post-Dolly, some people hatched sensational supposes about clone armies goose-stepping through foreign capitals, or ranches where people would raise clones to harvest organs. Less outlandishly, some feared that clones would be burdened by disease or deep molecular flaws. Cloning adult DNA requires turning on dormant genes and pushing cells to divide, divide, divide. That sounds a lot like cancer, and clones do seem prone to tumors. Many scientists also concluded (although Dolly's midwives dispute this) that Dolly was born a genetic geriatric, with unnaturally old and decrepit cells. Arthritis did in fact stiffen Dolly's legs at a precocious age, and she died at age six (half her breed's life span) after contracting a virus that, à la Peyton Rous, gave her lung cancer. The adult DNA used to clone Dolly had been—like all adult DNA—pockmarked with epigenetic changes and warped by mutations and poorly patched breaks. Such flaws might have corrupted her genome before she was ever born.*

But if we're toying with playing god here, we might as well play devil's advocate, too. Suppose that scientists overcome all the medical limitations and produce perfectly healthy clones. Many people would still oppose human cloning on principle. Part of their reasoning, however, relies on understandable but thankfully faulty assumptions about genetic determinism, the idea that DNA rigidly dictates our biology and personality. With every new genome that scientists sequence, it becomes clearer that genes deal in probabilities, not certainties. A genetic *influence* is just that, only that. Just as important, epigenetic research shows that the environment changes how genes work and interact, so cloning someone faithfully might require preserving every epigenetic tag from every missed meal and every cigarette. (Good luck.) Most people forget too that it's already too late to avoid exposure to human clones; they live among us even now, monstrosities called identical twins. A clone and its parent would be no more alike than twins are with all their epigenetic differences, and there's reason to believe they'd actually be less alike.

Consider: Greek philosophers debated the idea of a ship whose hull and decks were gradually rotting, plank by plank; eventually, over the decades, every original scrap of wood got replaced. Was it still the same ship at the end? Why or why not? Human beings present a similar stumper. Atoms in our body get recycled many, many times before death, so we don't have the same bodies our whole lives. Nevertheless we feel like the same person. Why? Because unlike a ship, each human has an uninterrupted store of thoughts and remembrances. If the human soul exists, that mental memory cache is it. But a clone would have different memories than his parent—would grow up with different music and heroes, be exposed to different foods and chemicals, have a brain wired differently by new technologies. The sum of these differences would be dissimilar tastes and inclinations—leading to a dissimilar temperament and a distinct soul. Cloning would therefore not produce a doppelgänger in anything but literal superficialities. Our DNA does circumscribe us; but where we fall within our range of possibilities—our statures, what diseases we'll catch, how our brains handle stress or temptation or setbacks—depends on more than DNA.

Make no mistake, I'm not arguing in favor of cloning here. If anything, this argues against—since what would be the point? Bereaved parents might yearn to clone Junior and ease that ache every time they walked by his empty room, or psychologists might want to clone Ted Kaczynski or Jim Jones and learn how to defuse sociopaths. But if cloning won't fulfill those demands—and it almost certainly cannot—why bother?

Cloning not only riles people up over unlikely horrors, it distracts from other controversies about human nature that genetic research can, and has, dredged up. As much as we'd like to close our eyes to these quarrels, they don't seem likely to vanish.

Sexual orientation has some genetic basis. Bees, birds, beetles, crabs, fish, skinks, snakes, toads, and mammals of all stripes (bison, lions, raccoons, dolphins, bears, monkeys) happily get frisky with their own sex, and their coupling often seems hardwired. Scientists have discovered that disabling even a single gene in mice—the suggestively named *fucM* gene—can turn female mice into lesbians. Human sexuality is more nuanced, but gay men (who have been studied more extensively than gay women) have substantially more gay relatives than heterosexual men raised in similar circumstances, and genes seem like one strong differentiator.

This presents a Darwinian conundrum. Being gay decreases the likelihood of having children and passing on any "gay genes," yet homosexuality has persisted in every last corner of the globe throughout all of history, despite often-violent persecution. One theory argues that perhaps gay genes are really "man-loving" genes—androphilic DNA that makes men love men but also makes women who have it lust after men, too, increasing their odds of having children. (Vice versa for gynophilic DNA.) Or perhaps homosexuality arises as a side effect of other genetic interactions. Multiple studies have found higher rates of left-handedness and ambidextrousness among gay men, and gay men frequently have longer ring fingers, too. No one really believes that holding a salad fork in one hand or the other causes homosexuality, but some far-reaching gene might influence both traits, perhaps by fiddling with the brain.

These discoveries are doubled-edged. Finding genetic links would validate being gay as innate and intrinsic, not a deviant "choice." That said, people already tremble about the possibility of screening for and singling out homosexuals, even potential homosexuals, from a young age. What's more, these results can be misrepresented. One strong predictor of homosexuality is the number of older biological brothers someone has; each one increases the odds by 20 to 30 percent. The leading explanation is that a mother's immune system mounts a progressively stronger response to each "foreign" Y chromosome in her uterus, and this immune response somehow induces homosexuality in the fetal brain. Again, this would ground homosexuality in biology— but you can see how a naive, or malicious, observer could twist this immunity link rhetorically and equate homosexuality with a disease to eradicate. It's a fraught picture.

Race also causes a lot of discomfort among geneticists. For one thing, the existence of races makes little sense. Humans have lower genetic diversity than almost any animal, but our colors and proportions and facial features vary as wildly as the finalists each year at Crufts. One theory of race argues that near extinctions isolated pockets of early humans with slight variations, and as these groups migrated beyond Africa and bred with Neanderthals and Denisovans and who knows what else, those variations became exaggerated. Regardless, some DNA must differ between ethnic groups: an aboriginal Australian husband and wife will never themselves produce a freckled, red-haired Seamus, even if they move to the Emerald Isle and breed till doomsday. Color is encoded in DNA.

The sticking point, obviously, isn't Maybelline-like variations in skin tone but other potential differences. Bruce Lahn, a geneticist at the University of Chicago, started his career cataloging palindromes and inversions on Y chromosomes, but around 2005 he began studying the brain genes *microcephalin* and *aspm*, which influence the growth of neurons. Although multiple versions exist in humans, one version of each gene had numerous hitchhikers and seemed to have swept through our ancestors at about Mach 10. This implied a strong survival advantage, and based on their ability to grow neurons, Lahn took a small leap and argued that these genes gave humans a cognitive boost. Intriguingly, he noted that the brain-boosting versions of *microcephalin* and *aspm* started to spread, respectively, around 35,000 BC and 4,000 BC, when, respectively, the first symbolic art and the first cities appeared in history. Hot on the trail, Lahn screened different populations alive today and determined that the brain-boosting versions appeared several times more often among Asians and Caucasians than among native Africans. *Gulp.*

Other scientists denounced the findings as speculative, irresponsible, racist, and wrong. These two genes exercise themselves in many places beyond the brain, so they may have aided ancient Europeans and Asians in other ways. The genes seem to help sperm whip their tails faster, for one thing, and might have outfitted the immune system with new weapons. (They've also been linked to perfect pitch, as well as tonal languages.) Even more damning, follow-up studies determined that people with these genes scored no better on IQ tests than those without them. This pretty much killed the brain-boosting hypothesis, and Lahn—who, for what it's worth, is a Chinese immigrant— soon admitted, "On the scientific level, I am a little bit disap- pointed. But in the context of the social and political controversy, I am a little bit relieved."

He wasn't the only one: race really bifurcates geneticists. Some swear up and down that race doesn't exist. It's "biologi- cally meaningless," they maintain, a social construct. *Race* is indeed a loaded term, and most geneticists prefer to speak some- what euphemistically of "ethnic groups" or "populations," which they confess do exist. But even then some geneticists want to censor investigations into ethnic groups and mental aptitude as inherently wounding—they want a moratorium. Others remain confident that any good study will just prove racial equality, so what the hey, let them continue. (Of course the act of lecturing us about race, even to point out its nonexistence, probably just reinforces the idea. Quick—don't think of green giraffes.)

Meanwhile some otherwise very pious scientists think the "biologically meaningless" bit is baloney. For one thing, some ethnic groups respond poorly—for purely biochemical reasons— to certain medications for hepatitis C and heart disease, among other ailments. Other groups, because of meager conditions in their ancient homelands, have become vulnerable to metabolic disorders in modern times of plenty. One controversial theory argues that descendants of people captured in slave raids in Africa have elevated rates of hypertension today in part because ancestors of theirs whose bodies hoarded nutrients, especially salt, more easily survived the awful oceanic voyages to their new homes. A few ethnic groups even have higher immunity to HIV, but each group, again, for different biochemical reasons. In these and other cases—Crohn's disease, diabetes, breast cancer—doctors and epidemiologists who deny race completely could harm people.

On a broader level, some scientists argue that races exist because each geographic population has, indisputably, distinct versions of some genes. If you examine even a few hundred snippets of someone's DNA, you can segregate him into one of a few broad ancestral groups nearly 100 percent of the time. Like it or not, those groups do generally correspond to people's traditional notion of races—African, Asian, Caucasian (or "swine-pink," as one anthropologist put it), and so on. True, there's always genetic bleed-over between ethnic groups, especially at geographic crossroads like India, a fact that renders the concept of race useless—too imprecise—for many scientific studies. But people's self-identified social race does predict their biological population group pretty well. And because we don't know what every distinct version of every stretch of DNA does, a few polemical and very stubborn scientists who study races/populations/whatever-you-want-to-call-thems argue that exploring potential differences in intellect is fair game—they resent being censored. Predictably, both those who affirm and those who deny race accuse the other side of letting politics color their science.*

Beyond race and sexuality, genetics has popped up recently in discussions of crime, gender relations, addiction, obesity, and many other things. Over the next few decades, in fact, genetic factors and susceptibilities will probably emerge for almost every human trait or behavior—take the over on that one. But regardless of what geneticists discover about these traits or behaviors, we should keep a few guidelines in mind when applying genetics to social issues. Most important, no matter the biological under-pinnings of a trait, ask yourself if it really makes sense to con-demn or dismiss someone based on how a few microscopic genes behave. Also, remember that most of our genetic predilections for behavior were shaped by the African savanna many thou-sands if not millions of years ago. So while "natural" in some sense, these predilections don't necessarily serve us well today, since we live in a radically different environment. What happens in nature is a poor guide for making decisions anyway. One of the biggest boners in ethical philosophy is the naturalistic fal-lacy, which equates nature with "what's right" and uses "what's natural" to justify or excuse prejudice. We human beings are *humane* in part because we can look beyond our biology.

In any study that touches on social issues, we can at least pause and not draw sensational conclusions without reasonably complete evidence. In the past five years, scientists have consci-entiously sought out and sequenced DNA from more and more ethnic groups worldwide, to expand what remains, even today, an overwhelmingly European pool of genomes available to study. And some early results, especially from the self-explanatory 1,000 Genomes Project, indicate that scientists might have over-estimated the importance of genetic sweeps—the same sweeps that ignited Lahn's race-intelligence firecracker.

By 2010 geneticists had identified two thousand versions of human genes that showed signs of being swept along; specifi-cally, because of low diversity around these genes, it looked as if hitchhiking had taken place. And when scientists looked for what differentiated these swept-along versions from versions not swept along, they found cases where a DNA triplet had mutated and now called for a new amino acid. This made sense: a new amino acid could change the protein, and if that change made someone fitter, natural selection might indeed sweep it through a population. However, when scientists examined other regions, they found the same signs of sweeps in genes with *silent* mutations—mutations that, because of redundancy in the genetic code, didn't change the amino acid. Natural selection cannot have swept these changes along, because the mutation would be invisible and offer no benefits. In other words, many apparent DNA sweeps could be spurious, artifacts of other evolutionary processes.

That doesn't mean that sweeps never happen; scientists still believe that genes for lactose tolerance, hair structure, and a few other traits (including, ironically, skin color) did sweep through various ethnic groups at various points as migrants encountered new environments beyond Africa. But those might represent rare cases. Most human changes spread slowly, and probably no one ethnic group ever "leaped ahead" in a genetic sweepstakes by acquiring blockbuster genes. Any claims to the contrary— especially considering how often supposedly scientific claims about ethnic groups have fallen apart before—should be handled with caution. Because as the old saw says, it's not what we don't know that stirs up trouble, it's what we do know that just ain't so.

Becoming wiser in the ways of genetics will require not only advances in understanding how genes work, but advances in computing power. Moore's Law for computers—which says that microchips get roughly twice as powerful every two years—has held for decades, which explains why some pet collars today could outperform the Apollo mission mainframes. But since 1990 genetic technology has outstripped even Moore's projections. A modern DNA sequencer can generate more data in twenty-four hours than the Human Genome Project did in ten long years, and the technology has become increasingly convenient, spreading to labs and field stations worldwide. (After killing Osama bin Laden in 2011, U.S. military personnel identified him—by matching his DNA to samples collected from relatives—within hours, in the middle of the ocean, in the dead of the a.m.) Simultaneously, the cost of sequencing an entire genome has gone into vacuum free-fall—from $3,000,000,000 to $10,000, from $1 per base pair to around 0.0003¢. If scientists want to study a single gene nowadays, it's often cheaper to sequence the entire genome instead of bothering to isolate the gene first and sequence just that part.

Of course, scientists still need to *analyze* the bajillions of A's, C's, G's, and T's they're gathering. Having been humbled by the HGP, they know they can't just stare at the stream of raw data and expect insights to pop out, *Matrix* style. They need to consider how cells splice DNA and add epigenetic marginalia, much more complicated processes. They need to study how genes work in groups and how DNA packages itself in three dimensions inside the nucleus. Equally important, they need to determine how culture—itself a partial product of DNA—bends back and influences genetic evolution. Indeed, some scientists argue that the feedback loop between DNA and culture has not only influenced but outright dominated human evolution over the past sixty thousand years or so. Getting a handle on all of this will require serious computing horsepower. Craig Venter demanded a supercomputer, but geneticists in the future might need to turn to DNA itself, and develop tools based on its amazing computational powers.

On the software side of things, so-called genetic algorithms can help solve complicated problems by harnessing the power of evolution. In short, genetic algorithms treat the computer commands that programmers string together as individual "genes" strung together to make digital "chromosomes." The programmer might start with a dozen different programs to test. He encodes the gene-commands in each one as binary 0s and 1s and strings them together into one long, chromosome-like sequence (0001010111011101010...). Then comes the fun part. The programmer runs each program, evaluates it, and orders the best programs to "cross over"—to exchange strings of 0s and 1s, just like chromosomes exchange DNA. Next the programmer runs these hybrid programs and evaluates them. At this point the best cross over and exchange more 0s and 1s. The process then repeats, and continues again, and again, allowing the programs to evolve. Occasional "mutations"—flipping 0s to 1s, or vice versa—add more variety. Overall, genetic algorithms combine the best "genes" of many different programs into one near-optimal one. Even if you start with moronic programs, genetic evolution improves them automatically and zooms in on better ones.

On the hardware (or "wetware") side of things, DNA could someday replace or augment silicon transistors and physically perform calculations. In one famous demonstration, a scientist used DNA to solve the classic traveling salesman problem. (In this brainteaser, a salesman has to travel to, say, eight cities scattered all over a map. He must visit each city once, but once he leaves a city he cannot visit it again, even just to pass through on his way somewhere else. Unfortunately, the cities have convoluted roads between them, so it's not obvious in what order to visit.)

To see how DNA could possibly solve this problem, consider a hypothetical example. First thing, you'd make two sets of DNA snippets. All are single-stranded. The first set consists of the eight cities to visit, and these snippets can be random A-C-G-T strings: Sioux Falls might be AGCTACAT, Kalamazoo TCGA-CAAT. For the second set, use the map. Every road between two cities gets a DNA snippet. However—here's the key—instead of making these snippets random, you do something clever. Say Highway 1 starts in Sioux Falls and ends in Kalamazoo. If you make the first half of the highway's snippet the A/T and C/G complement of half of Sioux Falls's letters, and make the second half of the highway's snippet the A/T and C/G complement of half of Kalamazoo's letters, then Highway 1's snippet can physically link the two cities:

```
(Sioux Falls)        (Kalamazoo)           (Fargo)

AGCTACAT      TCGACAAT        GTAGTAAT...

      \ \ \ \   / / / /   \ \ \ \   / / / /

      TGTAAGCT        GTTACATC...
        (Road 1)              (Road 2)
```

After encoding every other road and city in a similar way, the calculation begins. You mix a pinch of all these DNA snippets in a test tube, and presto change-o, one good shake computes the answer: somewhere in the vial will be a longer string of (now) double-stranded DNA, with the eight cities along one strand, in the order the salesman should visit, and all the connecting roads on the complementary strand.

Of course, that answer will be written in the biological equivalent of machine code (GCGAGACGTACGAATCC...) and will need deciphering. And while the test tube contains many copies of the correct answer, free-floating DNA is unruly, and the tube also contains trillions of wrong solutions—solutions that skipped cities or looped back and forth endlessly between two cities. Moreover, isolating the answer requires a tedious week of purifying the right DNA string in the lab. So, yeah, DNA computing isn't ready for *Jeopardy*. Still, you can understand the buzz. One gram of DNA can store the equivalent of a trillion CDs, which makes our laptops look like the gymnasium-sized behemoths of yesteryear. Plus, these "DNA transistors" can work on multiple calculations simultaneously much more easily than silicon circuits. Perhaps best of all, DNA transistors can assemble and copy themselves at little cost.

If deoxyribonucleic acid can indeed replace silicon in computers, geneticists would effectively be using DNA to analyze its own habits and history. DNA can already recognize itself; that's how its strands bond together. So DNA computers would give the molecule another modest level of reflexivity and self-awareness. DNA computers could even help DNA refine itself and improve its own function. (Makes you wonder who's in charge...)

And what kinds of DNA improvements might DNA computing bring about? Most obviously, we could eradicate the subtle malfunctions and stutters that lead to many genetic diseases. This controlled evolution would finally allow us to circumvent the grim waste of natural selection, which requires that the many be born with genetic flaws simply so that the few can advance incrementally. We might also improve our daily health, cinching our stomachs in by engineering a gene to burn high-fructose corn syrup (a modern answer to the ancient *apoE* meat-eating gene). More wildly, we could possibly reprogram our fingerprints or hairstyles. If global temperatures climb and climb, we might want to increase our surface area somehow to radiate heat, since squatter bodies retain more heat. (There's a reason Neanderthals in Ice Age Europe had beer keg chests.) Furthermore, some thinkers suggest making DNA adjustments not by tweaking existing genes but by putting updates on an extra pair of chromosomes and inserting them into embryos*—a software patch. This might prevent intergenerational breeding but would bring us back in line with the primate norm of forty-eight.

These changes could make human DNA worldwide even more alike than it is now. If we tinker with our hair and eye color and figures, we might end up looking alike, too. But based on the historical pattern with other technologies, things might well go the other way instead: our DNA could become as diverse as our taste in clothing, music, and food. In that case, DNA could go all postmodern on us, and the very notion of a standard human genome could disappear. The genomic text would become a palimpsest, endlessly overwriteable, and the metaphor of DNA as "the" blueprint or "the" book of life would no longer hold.

Not that it ever really did hold, outside our imaginations. Unlike books and blueprints, both human creations, DNA has no fixed or deliberate meaning. Or rather, it has only the meaning we infuse it with. For this reason we should interpret DNA cautiously, less like prose, more like the complicated and solemn utterances of an oracle.

As with scientists studying DNA, pilgrims to the Delphic oracle in ancient Greece always learned something profound about themselves when they inquired of it—but rarely what they assumed they'd learned at first. The general-king Croesus once asked Delphi if he should engage another emperor in battle. The oracle answered, "You will destroy a great empire." Croesus did—his own. The oracle informed Socrates that "no one is wiser." Socrates doubted this, until he'd canvassed and interrogated all the reputedly wise men around. He then realized that, unlike them, he at least admitted his ignorance and didn't fool himself into "knowing" things he didn't. In both cases, the truth emerged only with time, with reflection, when people had gathered all the facts and could parse the ambiguities. The same with DNA: it all too often tells us what we want to hear, and any dramatist could learn a thing about irony from it.

Unlike Delphi, our oracle still speaks. From so humble a beginning, despite swerves and near extinctions, our DNA (and RNA and other 'NAs) did manage to create us—creatures bright enough to discover and decipher the DNA inside them. But bright enough as well to realize how much that DNA limits them. DNA has revealed a trove of stories about our past that we thought we'd lost forever, and it has endowed us with sufficient brains and curiosity to keep mining that trove for centuries more. And despite that push-pull, gotta-have-it-won't-stand-for-it ambivalence, the more we learn, the more tempting, even desirable, it seems to change that DNA. DNA endowed us with imagination, and we can now imagine freeing ourselves from the hard and heartbreaking shackles it puts on life. We can imagine remaking our very chemical essences; we can imagine remaking life as we know it. This oracular molecule seems to promise that if we just keep pushing, keep exploring and sounding out and tinkering with our genetic material, then life as we know it will cease. And beyond all the intrinsic beauty of genetics and all the sobering insights and all the unexpected laughs that it provides, it's that promise that keeps drawing us back to it, to learn more and more and yet still more about our DNA and our genes, our genes and our DNA.

Epilogue

Genomics Gets Personal

lthough they know better, many people versed in science, even many scientists, still fear their genes on some subliminal level. Because no matter how well you understand things intellectually, and no matter how many counterexamples turn up, it's still hard to accept that having a DNA signature for a disease doesn't condemn you to develop the disease itself. Even when this registers in the brain, the gut resists. This discord explains why memories of his Alzheimer's-ridden grandmother convinced James Watson to suppress his *apoE* status. It also explains, when I plumbed my own genes, why boyhood memories of fleeing from my grandfather convinced me to conceal any hints about Parkinson's disease.

During the writing of this book, however, I discovered that Craig Venter had published everything about his genome, uncensored. Even if releasing it publicly seemed foolhardy, I admired his aplomb in facing down his DNA. His example fortified me, and every day that passed, the discrepancy between what I'd concluded (that people should indeed face down their genes) and how I was behaving (hiding from my Parkinson's status) nagged me more and more. So eventually I sucked it up, logged on to the testing company, and clicked to break the electronic seal on that result.

Admittedly, it took another few seconds before I could look up from my lap to the screen. As soon as I did, I felt a narcotic of relief flood through me. I felt my shoulders and limbs unwind: according to the company, I had no increased risk for Parkinson's after all.

I whooped. I rejoiced—but should I have? There was a definite irony in my happiness. Genes don't deal in certainties; they deal in probabilities. That was my mantra before I peeked, my way of convincing myself that even if I had the risky DNA, it wouldn't *inevitably* ravage my brain. But when things looked less grim suddenly, I happily dispensed with uncertainty, happily ignored the fact that lower-risk DNA doesn't mean I've inevitably escaped anything. Genes deal in probabilities, and some probability still existed. I knew this—and for all that, my relief was no less real. It's the paradox of personal genetics.

Over the next months, I shooed away this inconvenient little cognitive dissonance and concentrated on finishing the book, forgetting that DNA always gets the last word. On the day I dotted the last *i*, the testing company announced some updates to old results, based on new scientific studies. I pulled up my browser and started scrolling. I'd seen previous rounds of updates before, and in each case the new results had merely corroborated what I'd already learned; my risks for things had certainly never changed much. So I barely hesitated when I saw an update for Parkinson's. Fortified and foolhardy, I clicked right through.

Before my mind registered anything, my eyes lit on some green letters in a large font, which reinforced my complacency. (Only red lettering would have meant *watch out*.) So I had to read the accompanying text a few times before I grasped it: "Slightly higher odds of developing Parkinson's disease."

Higher? I looked closer. A new study had scrutinized DNA at a different spot in the genome from the results I'd seen before. Most Caucasian people like me have either CT or TT at the spot in question, on chromosome four. I had (per the fat green letters) CC there. Which meant, said the study, higher odds.

I'd been double-crossed. To expect a genetic condemnation and receive it in due course is one thing. But to expect a condemnation, get pardoned, and find myself condemned again? Infinitely more torture.

Somehow, though, receiving this genetic sentence didn't tighten my throat as it should have. I felt no panic, either, no fight-or-flight jolt of neurotransmitters. Psychologically, this should have been the worst possible thing to endure—and yet my mind hadn't erupted. I wasn't exactly pumped up about the news, but I felt more or less tranquil, untroubled.

So what happened between the first revelation and the second, the setup and the would-be knockdown? Without sounding too pompous, I guess I got an education. I knew now that for a complex disease like Parkinson's—subject to the sway of many genes—any one gene probably contributes little to my risk. I then investigated what a "slightly higher" risk meant, anyway—just 20 percent, it turns out. And that's for a disease that affects (as further digging revealed) only 1.6 percent of men anyway. The new study was also, the company admitted, "preliminary," subject to amendments and perhaps outright reversals. I might still be saddled with Parkinson's as an old man; but somewhere in the generational shuffling of genes, somewhere between Grandpa Kean and Gene and Jean, the dangerous bits might well have been dealt out—and even if they're still lurking, there's no guarantee they'll flare up. There's no reason for the little boy in me to keep fleeing.

It had finally penetrated my skull: probabilities, not certainties. I'm not saying personal genetics is useless. I'm glad to know, for instance (as other studies tell me), that I face higher odds of developing prostate cancer, so I can always make sure the doctor dons a rubber glove to check for that as I age. (Something to look forward to.) But in the clinic, for a patient, genes are just another tool, like blood work or urinalysis or family history. Indeed, the most profound changes that genetic science brings about likely won't be instant diagnoses or medicinal panaceas but mental and spiritual enrichment—a more expansive sense of who we humans are, existentially, and how we fit with other life on earth. I enjoyed having my DNA sequenced and would do it again, but not because I might gain a health advantage. It's more that I'm glad I was here, am here, in the beginning.

ACKNOWLEDGMENTS

First off, a thank-you to my loved ones. To Paula, who once again held my hand and laughed with me (and at me when I deserved it). To my two siblings, two of the finest people around, lucky additions to my life. To all my other friends and family in D.C. and South Dakota and around the country, who helped me keep perspective. And finally to Gene and Jean, whose genes made this book possible. :)

I would furthermore like to thank my agent, Rick Broadhead, for embarking on another great book with me. And thank you as well to my editor at Little, Brown, John Parsley, who helped shape and improve the book immensely. Also invaluable were others at and around Little, Brown who've worked with me on this book and on *Spoon*, including William Boggess, Carolyn O'Keefe, Morgan Moroney, Peggy Freudenthal, Bill Henry, Deborah Jacobs, Katie Gehron, and many others. I offer thanks, too, to the many, many scientists and historians who contributed to individual chapters and passages, either by fleshing out stories, helping me hunt down information, or offering their time to explain something. If I've left anyone off this list, my apologies. I remain thankful, if embarrassed.

Notes and Errata

Chapter 1: *Genes, Freaks, DNA*

p. 26, The 3:1 ratio: Welcome to the endnotes! Wherever you see an asterisk (*) in the text, you can flip back here to find digressions, discussions, scuttlebutt, and errata about the subject at hand. If you want to flip back immediately for each note, go right ahead; or if you prefer, you can wait and read all the notes after finishing each chapter, as a sort of afterword. This first endnote provides a refresher for Mendelian ratios, so if you're comfortable with that, feel free to move along. But do flip back again. The notes get more salacious. Promise.

The refresher: Mendel worked with dominant traits (like tallness, capital *A*) and recessive traits (like shortness, lowercase *a*). Any plant or animal has two copies of each gene, one from Mom, one from Dad. So when Mendel crossed *AA* plants with *aa* plants (below, left), the progeny were all *Aa* and therefore all tall (since *A* dominates *a*):

	A	A
a	Aa	Aa
a	Aa	Aa

	A	a
A	AA	Aa
a	Aa	aa

When Mendel crossed one *Aa* with another (above, right), things got more interesting. Each *Aa* can pass down *A* or *a*, so there are four possibilities for the offspring: *AA*, *Aa*, *aA*, and *aa*. The first three are again tall, but the last one will be short, though it came from tall parents. Hence a 3:1 ratio. And just to be clear, the ratio holds in plants, animals, whatever; there's nothing special about peas.

The other standard Mendelian ratio comes about when *Aa* mates with *aa*. In this case, half the children will be *aa* and won't show the dominant trait. Half will be *Aa* and will show it.

A	a
a	Aa	aa
a	Aa	aa

This 1:1 pattern is especially common in family trees when a dominant *A* trait is rare or arises spontaneously through a mutation, since every rare *Aa* would have to mate with the more common *aa*.

Overall the 3:1 and 1:1 ratios pop up again and again in classic genetics. If you're curious, scientists identified the first recessive gene in humans in 1902, for a disorder that turned urine black. Three years later, they pinned down the first dominant human gene, for excessively stubby fingers.

Chapter 2: *The Near Death of Darwin*

p. 32, until the situation blew over: The details about Bridges's private life appear in *Lords of the Fly*, by Robert Kohler.

p. 34, to proceed by jumps: When they were both young, in the 1830s, Darwin convinced his first cousin, Francis Galton, to drop out of medical school and take up mathematics instead. Darwin's later defenders must have rued this advice many, many times, for it was Galton's pioneering statistical work on bell curves—and Galton's relentless arguments, based on that work—that most seriously undermined Darwin's reputation.

As detailed in *A Guinea Pig's History of Biology*, Galton had gathered some of his evidence for bell curves in his typically eccentric way, at the International Health Exhibition in London in 1884. The expo was as much a social affair as a scientific endeavor: as patrons wandered through exhibits about sanitation and sewers, they gulped mint juleps and spiked arrack punch and kumiss (fermented mare's milk produced by horses on-site) and generally had a gay old time. Galton set up a booth at the expo as well and doggedly measured the stature, eyesight, and hearing of nine thousand occasionally intoxicated Englishmen. He also tested their strength with fairground games that involved punching and squeezing various contraptions, a task that proved more difficult than Galton had anticipated: oafs who didn't understand the equipment constantly broke it, and others wanted to show off their strength and impress girls. It was a true fairground atmosphere, but Galton had little fun: he later described "the stupidity and wrong-headedness" of his fellow fairgoers as "so great as to be scarcely credible." But as expected, Galton gathered enough data to confirm that human traits also formed bell curves. The finding further bolstered his confidence that he, not Cousin Charles, understood how evolution proceeded, and that small variations and small changes played no important role.

This wasn't the first time Galton had thwarted Darwin, either. From the day he published *On the Origin of Species*, Darwin was aware that his theory lacked something, badly. Evolution by natural selection requires creatures to inherit favorable traits, but no one (save an obscure monk) had any idea how that worked. So Darwin spent his last years devising a theory, pangenesis, to explain that process.

Pangenesis held that each organ and limb pumped out microscopic spores called gemmules. These circulated inside a creature, carrying information about both its inborn traits (its nature) and also any traits it acquired during its lifetime (its environment, or nurture). These gemmules got filtered out by the body's erogenous zones, and copulation allowed male and female gemmules to mix like two drops of water when males deposited their semen.

Although ultimately mistaken, pangenesis was an elegant theory. So when Galton designed an equally elegant experiment to hunt for gemmules in rabbits, Darwin heartily encouraged the emprise. His hopes were soon dashed. Galton reasoned that if gemmules circulated, they must do so in the blood. So he began transfusing blood among black, white, and silver hares, hoping to produce a few mottled mongrels when they had children. But after years of breeding, the results were pretty black-and-white: not a single multishaded rabbit appeared. Galton published a quickie scientific paper suggesting that gemmules didn't exist, at which point the normally avuncular Darwin went apoplectic. The two men had been warmly exchanging letters for years on scientific and personal topics, often flattering each other's ideas. This time Darwin lit into Galton, fuming that he'd never once mentioned gemmules circulating in blood, so transfusing blood among rabbits didn't prove a damn thing.

On top of being disingenuous—Darwin hadn't said boo about blood not being a good vehicle for gemmules when Galton was doing all the work—Darwin was deceiving himself here. Galton had indeed destroyed pangenesis and gemmules in one blow.

p. 42, together on one chromosome: Sex-linked recessive traits like these show up more often in males than in females for a simple reason. An XX female with a rare white-eye gene on one X would almost certainly have the red-eye gene on the other X. Since red dominates white, she wouldn't have white eyes. But an XY male has no backup if he gets the white-eye gene on his X; he will be white-eyed by default. Geneticists call females with one recessive version "carriers," and they pass the gene to half their male children. In humans hemophilia is one example of a sex-linked trait, Sturtevant's red-green color blindness another.

p. 47, produce millions of descendants: Many different books talk a bit about the fly room, but for the full history, check out *A Guinea Pig's History of Biology*, by Jim Endersby, one of my favorite books ever. Endersby also touches on Darwin's adventures with gemmules, Barbara McClintock (from chapter 5), and other fascinating tales.

p. 49, his reputation would never lapse again: A historian once wisely noted that "in reading Darwin, as in reading Shakespeare or the Bible, it is possible to support almost any viewpoint desirable by focusing on certain isolated passages." So you have to be careful when drawing broad conclusions from Darwin quotes. That said, Darwin's antipathy for math seemed genuine, and some have suggested that even elementary equations frustrated him. In one of history's ironies, Darwin ran his own experiments on plants in the primrose genus, just like de Vries, and came up with clear 3:1 ratios among offspring traits. He obviously wouldn't have linked this to Mendel, but he seems not to have grasped that ratios might be important at all.

p. 52, inside fruit fly spit glands: *Drosophila* go through a pupa stage where they encase themselves in gluey saliva. To get as many saliva-producing genes going as possible, salivary-gland cells repeatedly double their chromosomes, which creates gigantic "puff chromosomes," chromosomes of truly Brobdingnagian stature.

Chapter 3: *Them's the DNA Breaks*

p. 60, the "Central Dogma" of molecular biology: Despite its regal name, the Central Dogma has a mixed legacy. At first, Crick intended the dogma to mean something general like *DNA makes RNA, RNA makes proteins.* Later he reformulated it more precisely, talking about how "information" flowed from DNA to RNA to protein. But not every scientist absorbed the second iteration, and just like old-time religious dogmas, this one ended up shutting down rational thought among some adherents. "Dogma" implies unquestionable truth, and Crick later admitted, roaring with laughter, that he hadn't even known the definition of dogma when he defined his—it just sounded learned. Other scientists paid attention in church, however, and as word of this supposedly inviolable dogma spread, it transmogrified in many people's minds into something less precise, something more like *DNA exists just to make RNA, RNA just to make proteins.* Textbooks sometimes refer to this as the Central Dogma even today. Unfortunately this bastardized dogma seriously skews the truth. It hindered for decades (and still occasionally hinders) the recognition that DNA and especially RNA do much, much more than make proteins.

Indeed, while basic protein production requires messenger RNA (mRNA), transfer RNA (tRNA), and ribosomal RNA (rRNA), dozens of other kinds of regulatory RNA exist. Learning about all the different functions of RNA is like doing a crossword puzzle when you know the last letters of an answer but not the opening, and you run through the alphabet under your breath. I've seen references to aRNA, bRNA, cRNA, dRNA, eRNA, fRNA, and so on, even the scrabbulous qRNA and zRNA. There's also rasiRNA and tasiRNA, piRNA, snoRNA, the Steve Jobs–ish RNAi, and others. Thankfully, mRNA, rRNA, and tRNA cover all the genetics we'll need in this book.

p. 62, can represent the same amino acid: To clarify, each triplet represents only one amino acid. But the inverse is *not* true, because some amino acids are represented by more than one triplet. As an example, GGG can only be glycine. But GGU, GGC, and GGA also code for glycine, and that's where the redundancy comes in, because we really don't need all four.

p. 69, onto the succeeding generation: A few other events in history have exposed masses of people to radioactivity, most notoriously at the Chernobyl nuclear power plant in modern Ukraine. The 1986 Chernobyl meltdown exposed people to different types of radioactivity than the Hiroshima and Nagasaki bombs—fewer gamma rays and more radioactive versions of elements like cesium, strontium, and iodine, which can invade the body and unload on DNA at short range. Soviet officials compounded the problem by allowing crops to be harvested downwind of the accident and allowing cows to graze on exposed grass, then letting people eat and drink the contaminated milk and produce. The Chernobyl region has already reported some seven thousand cases of thyroid cancer, and medical officials expect sixteen thousand extra cancer deaths over the next few decades, an increase of 0.1 percent over background cancer levels.

And in contrast to Hiroshima and Nagasaki, the DNA of children of Chernobyl victims, especially children of men near Chernobyl, does show signs of increased mutations. These results remain disputed, but given the different exposure patterns and dosage levels—Chernobyl released hundreds of times more radioactivity than either atomic bomb—they could be real. Whether those mutations actually translate to long-term health problems among Chernobyl babies remains to be seen. (As an imperfect comparison, some plants and birds born after Chernobyl showed high mutation rates, but most seemed to suffer little for that.)

Sadly, Japan will now have to monitor its citizens once again for the long-term effects of fallout because of the breech of the Fukushima Daiichi nuclear power plant in spring 2011. Early government reports (some of which have been challenged) indicate that the damage was contained to an area one-tenth the size of Chernobyl's exposure footprint, mostly because radioactive elements at Chernobyl escaped into the air, while in Japan the ground and water absorbed them. Japan also intercepted most contaminated food and drink near Fukushima within six days. As a result, medical experts suspect the total number of cancer deaths in Japan will be correspondingly small—around one thousand extra deaths over the next few decades, compared to the twenty thousand who died in the earthquake and tsunami.

p. 71, just beginning to explore: For a full account of Yamaguchi's story—and for eight other equally riveting tales—see *Nine Who Survived Hiroshima and Nagasaki*, by Robert Trumbull. I can't recommend it highly enough.

For more detail on Muller and many other players in early genetics (including Thomas Hunt Morgan), check out the wonderfully comprehensive *Mendel's Legacy*, by Elof Axel Carlson, a former student of Muller's.

For a detailed but readable account of the physics, chemistry, and biology of how radioactive particles batter DNA, see *Radiobiology for the Radiologist*, by Eric J. Hall and Amato J. Giaccia. They also discuss the Hiroshima and Nagasaki bombs specifically.

Finally, for an entertaining rundown of early attempts to decipher the genetic code, I recommend Brian Hayes's "The Invention of the Genetic Code" in the January–February 1998 issue of *American Scientist*.

Chapter 4: *The Musical Score of DNA*

p. 78, even meant, if anything: Zipf himself believed that his law revealed something universal about the human mind: laziness. When speaking, we want to expend as little energy as possible getting our points across, he argued, so we use common words like *bad* because they're short and pop easily to mind. What prevents us from describing every last coward, rogue, scuzzbag, bastard, malcontent, coxcomb, shit-for-brains, and misanthrope as "bad" is other people's laziness, since they don't want to mentally parse every possible meaning of the word. They want precision, pronto. This tug-of-war of slothfulness results in languages where common words do the bulk of the work, but rarer and more descriptive words must appear now and then to appease the damn readers.

That's clever as far as it goes, but many researchers argue that any "deep" explanation of Zipf's law is, to use another common word, crap. They point out that something like a Zipfian distribution can arise in almost any chaotic situation. Even computer programs that spit out random letters and spaces—digital orangutans banging typewriters—can show Zipfian distributions in the resulting "words."

p. 82, evolution creeps forward: The analogy between genetic language and human language seems fuzzy to some, almost too cute to be true. Analogies can always be taken too far, but I think that some of this dismissal stems from our sort-of-selfish tendency to think that language can only be sounds that humans make. Language is wider than just us: it's the rules that govern any communication. And cells as surely as people can take feedback from their environment and adjust what they "say" in response. That they do so with molecules instead of air-pressure waves (i.e., sound) shouldn't prejudice us. Recognizing this, a few recent cellular biology textbooks have included chapters on Chomsky's theories about the underlying structure of languages.

p. 84, *sator…rotas:* The palindrome means something like "The farmer Arepo works with his plow," with *rotas*, literally "wheels," referring to the back-and-forth motion that plows make as they till. This "magic square" has delighted enigmatologists for centuries, but scholars have suggested it might have served another purpose during imperial Roman reigns of terror. An anagram of these twenty-five letters spells out *paternoster*, "Our Father," twice, in an interlocking cross. The four letters left over from the anagram, two *a*'s and two *o*'s, could then refer to the alpha and omega (famous later from the Book of Revelation). The theory is that, by sketching this innocuous palindrome on their doors, Christians could signal each other without arousing Roman suspicion. The magic square also reportedly kept away the devil, who traditionally (so said the church) got confused when he read palindromes.

p. 88, his boss's lunatic projects: Friedman's boss, "Colonel" George Fabyan, had quite a life. Fabyan's father started a cotton company called Bliss Fabyan and groomed Fabyan to take over. But succumbing to wanderlust, the boy ran away to work as a Minnesota lumberjack instead, and his outraged and betrayed father disinherited him. After two years, Fabyan tired of playing Paul Bunyan and decided to get back into the family business—by applying, under an assumed name, to a Bliss Fabyan office in St. Louis. He quickly set all sort of sales records, and his father at corporate HQ in Boston soon summoned this young go-getter to his office to talk about a promotion. In walked his son.

After this Shakespearean reunion, Fabyan thrived in the cotton business and used his wealth to open the think tank. He funded all sorts of research over the years but fixated on Shakespeare codes. He tried to publish a book after he supposedly broke the code, but a filmmaker working on some adaptations of Shakespeare sued to stop publication, arguing that its contents would "shatter" Shakespeare's reputation. For whatever reason, the local judge took the case—centuries of literary criticism apparently fell under his jurisdiction—and, incredibly, sided with Fabyan. His decision concluded, "Francis Bacon is the author of the works so erroneously attributed to William Shakespeare," and he ordered the film producer to pay Fabyan $5,000 in damages.

Most scholars look on arguments against Shakespeare's authorship about as kindly as biologists do on theories of maternal impressions. But several U.S. Supreme Court justices, most recently in 2009, have also voiced opinions that Shakespeare could not have written his plays. The real lesson here is that lawyers apparently have different standards of truth and evidence than scientists and historians.

p. 89, to rip off casinos at roulette: The casino gambit never paid off. The idea started with the engineer Edward Thorp, who in 1960 recruited Shannon to help him. At the roulette table, the two men worked as a team, though they pretended not to know each other. One watched the roulette ball as it spun around the wheel and noted the exact moment it passed certain points. He then used a toe-operated switch in his shoe to send signals to the small computer in his pocket, which in turn transmitted radio signals. The other man, wearing an earpiece, heard these signals as musical notes, and based on the tune, he would know where to toss his money. They painted any extruding wires (like the earpiece's) the color of flesh and pasted the wires to their skin with spirit gum.

Thorp and Shannon calculated an expected yield of 44 percent from their scheme, but Shannon turned chicken on their first test run in a casino and would only place dime bets. They won more often than not, but, perhaps after eyeing some of the heavies manning the casino door, Shannon lost his appetite for the enterprise. (Considering that the two men had ordered a $1,500 roulette wheel from Reno to practice, they probably *lost* money on the venture.) Abandoned by his partner, Thorp published his work, but it apparently took a number of years before casinos banned portable electronics outright.

Chapter 5: *DNA Vindication*

p. 97, an inside-out (and triple-stranded): For an account of the embarrassment and scorn Watson and Crick endured for this odd DNA model, please see my previous book, *The Disappearing Spoon*.

p. 102, complex and beautiful life: For a more detailed account of Miriam's life, I highly recommend *The Soul of DNA*, by Jun Tsuji.

p. 107, the oldest matrilineal ancestor: Using this logic, scientists also know that Mitochondrial Eve had a partner. All males inherit Y chromosomes from their fathers alone, since females lack the Y. So all men can trace strictly paternal lines back to find this Y-chromosomal Adam. The kicker is that, while simple laws of mathematics prove that this Adam and Eve must have existed, the same laws reveal that Eve lived tens of thousands of years earlier than Adam. So the Edenic couple could never have met, even if you take into account the extraordinary life expectancies in the Bible.

By the by, if we relax the strictly patrilineal or strictly matrilineal bit and look for the last ancestor who—through men *or* women—passed at least some DNA to every person alive today, that person lived only about five thousand years ago, long after humans had spread over the entire earth. Humans are strongly tribal, but genes always find a way to spread.

p. 114, got downgraded: Some historians argue that McClintock struggled to communicate her ideas partly because she couldn't draw, or at least didn't. By the 1950s molecular biologists and geneticists had developed highly stylized cartoon flowcharts to describe genetic processes. McClintock, from an older generation, never learned their drawing conventions, a deficit that—combined with the complexity of maize in the first place—might have made her ideas seem too convoluted. Indeed, some students of McClintock recall that they never remember her drawing any diagrams, ever, to explain anything. She was simply a verbal person, rutted in *logos*.

Compare this to Albert Einstein, who always maintained that he thought in pictures, even about the fundamentals of space and time. Charles Darwin was of McClintock's ilk. He included just one picture, of a tree of life, in the hundreds of pages of *On the Origin of Species*, and one historian who studied Darwin's original notebook sketches of plants and animals acknowledged he was a "terrible drawer."

p. 115, she withdrew from science: If you're interested in learning more about the reception of McClintock's work, the scholar most responsible for challenging the canonical version of her life's story is Nathaniel Comfort.

Chapter 6: *The Survivors, the Livers*

p. 125, producing the very Cyclops: Most children born with cyclopia (the medical term) don't live much past delivery. But a girl born with cyclopia in India in 2006 astounded doctors by surviving for at least two weeks, long enough for her parents to take her home. (No further information about her survival was available after the initial news reports.) Given the girl's classic symptoms—an undivided brain, no nose, and a single eye—it was almost certain that *sonic hedgehog* had malfunctioned. And sure enough, news outlets reported that the mother had taken an experimental cancer drug that blocks *sonic*.

p. 125, Maurice of Nassau: Prince Mo belonged to the dynastic House of Orange in the Netherlands, a family with an unusual (and possibly apocryphal) legend attached to its name. Centuries ago, wild carrots were predominantly purple. But right around 1600, Dutch carrot farmers, indulging in old-fashioned genetic engineering, began to breed and cultivate some mutants that happened to have high concentrations of the vitamin A variant beta carotene—and in doing so developed the first orange carrots. Whether farmers did this on their own or (as some historians claim) to honor Maurice's family remains unknown, but they forever changed the texture, flavor, and color of this vegetable.

p. 130, German biologist August Weismann: Although an undisputed brainiac and hall-of-fame biologist, Weismann once claimed—uproariously, given the book's mammoth size—to have read *On the Origin of Species* in one sitting.

p. 131, a fifth official letter to the DNAlphabet: A few scientists have even expanded the alphabet to six, seven, or eight letters, based on chemical variations of methylated cytosine. Those letters are called (if you're into the whole brevity thing) hmC, fC, and caC. It's not clear, though, whether these "letters" function independently or are just intermediate steps in the convoluted process by which cells strip the m from mC.

p. 136, and Arctic huskies: The tale of the husky liver is dramatic, involving a doomed expedition to reach the South Pole. I won't expand on the story here, but I have written something up and posted it online at http://samkean.com/thumb-notes. My website also contains links to tons of pictures (http://samkean.com/thumb-pictures), as well as other notes a little too digressive to include even here. So if you're interested in reading about Darwin's role in musicals, perusing an infamous scientific fraud's suicide note, or seeing painter Henri Toulouse-Lautrec nude on a public beach, take a look-see.

p. 136, carried the men home to the Netherlands: Europeans did not set eyes on the Huys again until 1871, when a party of explorers tracked it down. The white beams were green with lichen, and they found the hut sealed hermetically in ice. The explorers recovered, among other detritus, swords, books, a clock, a coin, utensils, "muskets, a flute, the small shoes of the ship's boy who had died there, and the letter Barents put up the chimney for safekeeping" to justify what some might see as a cowardly decision to abandon his ship on the ice.

Chapter 7: *The Machiavelli Microbe*

p. 141, the "RNA world" theory: Though RNA probably preceded DNA, other nucleic acids—like GNA, PNA, or TNA—might have preceded both of them. DNA builds its backbone from ringed deoxyribose sugars, which are more complicated than the building blocks likely available on the primordial earth. Glycol nucleic acid and peptide nucleic acid look like better candidates because neither uses ringed sugars for its vertebrae. (PNA doesn't use phosphates either.) Threose nucleic acid does use ringed sugars, but again, sugars simpler than DNA's. Scientists suspect those simpler backbones proved more robust as well, giving these 'NAs an advantage over DNA on the sun-scorched, semimolten, and oft-bombarded early earth.

p. 143, viruses that infect only other parasites: This idea of parasites feasting on parasites always puts me in mind of a wonderful bit of doggerel by Jonathan Swift:

> *So nat'ralists observe, a flea*
> *Hath smaller fleas that on him prey,*
> *And these have smaller fleas that bite 'em,*
> *And so proceed ad infinitum.*

For my taste, a mathematician named Augustus De Morgan outdid even Swift on this theme:

Great fleas have little fleas upon their backs to bite 'em,
And little fleas have lesser fleas, and so ad infinitum.
And the great fleas themselves, in turn, have greater fleas to go on,
While these again have greater still, and greater still, and so on.

p. 148, gave each cat an individual name: A sample: Stinky, Blindy, Sam, Pain-in-the-Ass, Fat Fuck, Pinky, Tom, Muffin, Tortoise, Stray, Pumpkin, Yankee, Yappy, Boots the First, Boots the Second, Boots the Third, Tigger, and Whisky.

p. 148, despite their mounting distress: In addition to the $111,000 yearly, there were occasional unexpected costs, like when an animal lib person cut a hole in the fence to spring as many cats as possible. Jack said there were still so many cats around that they didn't notice the dozens that escaped until a nun knocked on their door and asked if the cats climbing onto roofs throughout the neighborhood were theirs. Um, yes.

p. 150, a plausible biological basis for hoarding cats: To be scrupulous: scientists have not yet run controlled studies on the correlation between Toxo levels in the brain and hoarding. So it's possible that the link between Toxo, dopamine, cats, and hoarding could come to naught. Nor can Toxo explain everything about hoarding behavior, since people occasionally hoard dogs, too.

But most animal hoarders do hoard felines, and scientists involved in the Toxo studies find the link plausible and have said so publicly. They've simply seen too much evidence of how Toxo can change the hardwired behavior of rodents and other creatures. And regardless of how strong its influence turns out to be, Toxo still seeps dopamine into your brain.

p. 154, to help Jack cope: Over the years, Jack and Donna have given many interviews about their lives and struggles. A few sources include: *Cats I Have Known and Loved,* by Pierre Berton; "No Room to Swing a Cat!," by Philip Smith, *The People,* June 30, 1996; "Couple's Cat Colony Makes Record Books—and Lots of Work!," by Peter Cheney, *Toronto Star,* January 17, 1992; *Current Science,* August 31, 2001; "Kitty Fund," *Kitchener-Waterloo Record,* January 10, 1994; "$10,000 Averts Ruin for Owners of 633 Cats," by Kellie Hudson, *Toronto Star,* January 16, 1992; and *Scorned and Beloved: Dead of Winter Meetings with Canadian Eccentrics,* by Bill Richardson.

Chapter 8: *Love and Atavisms*

p. 160, customize proteins for different environments in the body: In one extreme example, fruit flies carve up the RNA of the *dscam* gene into 38,016 distinct products—roughly triple the number of genes a fruit fly has. So much for the one gene/one protein theory!

p. 162, practically a defining trait of mammals: Nature loves playing gotcha, and for almost everything you call a "unique" mammalian trait, there's an exception: reptiles with a rudimentary placenta, for instance, or insects that give birth to live young. But in general, these are mammalian traits.

p. 167, more extensive MHCs than other creatures: In humans, the MHC is often called the HLA, but since we're focused on mammals here, I'll use the general term.

p. 170, the same extra nipples that barnyard sows have: Though best known for the telephone, Alexander Graham Bell had a keen interest in genetics and dreamed of breeding fitter humans. To learn more about biology, he bred sheep with extra nipples and studied the inheritance patterns.

p. 172, tails contract involuntarily when children cough or sneeze: For more on human tails, see Jan Bondeson's wonderful *A Cabinet of Medical Curiosities*. The book also has an astounding chapter on maternal impressions (like those in chapter 1), as well as many other gruesome tales from the history of anatomy.

p. 174, simply have to round them up: The scientist did not end up winning funding for his research. And to be fair, he didn't intend to spend all $7.5 million developing the gay bomb. Some of that money would have gone to, among other projects, a separate bomb that would have given the enemy epically bad breath, to the point of inducing nausea. No word on whether the scientist ever realized he could combine the two bombs into the most frustrating weapon in history.

Chapter 9: *Humanzees and Other Near Misses*

p. 182, They're that similar: In fact, this is how scientists first determined that chimps, not gorillas, are our closest living relatives. Scientists performed the first DNA hybridization experiments in the 1980s by mixing chimp, gorilla, and human DNA in a hot, steamy bath. When things cooled down, human DNA stuck to chimp DNA more readily than it did to gorilla DNA. QED.

p. 186, should always have fewer mutations: This isn't the place to even attempt to resolve this debate, but the scientists who first proposed the interbreeding theory have of course tried to counterrefute this supposed refutation. And the original scientists do have a point: in their paper announcing the theory way back in 2006, they actually anticipated this criticism about the X looking more uniform because of sperm production rates. Specifically, they noted that while X chromosomes should indeed look more alike for that reason, the X chromosomes they studied looked even more alike than this scenario could account for.

Naturally, the refuting scientists are busy countering the counterrefutations. It's all very technical and a bit arcane, but exciting, given the stakes...

p. 189, The *Times* story: In addition to its salacious details, the *Times* story also included this bizarre—and bizarrely egalitarian—quote: one scientist was convinced "that if the orang[utan] be hybridized with the yellow race, the gorilla with the black race, and the chimpanzees with the white race, all three hybrids will reproduce themselves." What's striking, especially for the time, is the insistence that all human beings, regardless of color, were kin to brutes.

p. 196, the path to forty-six chromosomes a million years ago: To anticipate a question, yes, chromosomes can split, too, by a process called fission. In the primate line, our current chromosome numbers three and twenty-one were once yoked together into a team, and formed our longest chromosome for millions of years. Numbers fourteen and fifteen also split before the rise of great apes long ago, and both retain a funny, off-center shape today as a legacy. In some ways, then, the fourteen-fifteen fusion in the Chinese man was the ultimate genetic atavism, returning him to the ancestral, pre-ape state!

p. 198, months of trench warfare: For more on Ivanov's life, the most authoritative and least sensationalistic source is a paper by Kirill Rossiianov, in *Science in Context*, from the summer 2002 issue: "Beyond Species: Il'ya Ivanov and his experiments on cross-breeding humans with anthropoid apes."

Chapter 10: *Scarlet A's, C's, G's, and T's*

p. 204, "Until I tasted a bluebottle [fly]": The supply of Buckland anecdotes is pretty much bottomless. One of his friends' favorites was the time he and a stranger sitting across from him on a long train ride both fell asleep in their seats. Buckland woke up to find that some red slugs formerly nestled in his pockets had escaped, and were now sliming their way across his companion's bald pate. Buckland discreetly exited at the next stop. Buckland also inspired his equally eccentric son Frank, who inherited his predilection for zoophagy and actually pioneered some of the more outré dishes the Buckland family ate. Frank had a standing agreement with the London Zoo that he got a shank of whatever animals died there.

Despite insulting Buckland, Darwin also indulged in zoophagy, even joining the Glutton Club at Cambridge, where he and companions dined on hawks, owls, and other beasts. On the *Beagle* voyage, Darwin ate ostrich omelets and armadillo roasted in its case, and after tucking into an agouti, a coffee-colored rodent that weighs twenty pounds, he declared it "the very best meat I ever tasted."

For more details on Buckland's life, work, family, and eccentricities, I highly recommend *The Heyday of Natural History*, by Lynn Barber, and *Bones and Ochre*, by Marianne Sommer.

p. 209, He named it *Megalosaurus*: It later came to light that another scientist had discovered *Megalosaurus* bones in the 1600s, including a tree trunk of a femur. But he'd classified them as the bones of giant humans, a decision that Buckland's work disproved. Strangely, two knobs on the end of that femur apparently traced out with Michelangelo-like verisimilitude the lower half of the human male package, inspiring a less-than-dignified moniker for the purported giants. Arguably, then, based on scientific priority in naming, the first known dinosaur species should be called *Scrotum humanum*. Buckland's more proper name stuck, though.

p. 211, thick, glowering brow we still associate with Neanderthals: The professor who identified the purported Cossack had decided the brow was shaped that way because the victim spent so many days furrowing it in pain. The professor apparently even believed the Cossack had scampered up sixty feet of sheer rock while mortally wounded, disrobed completely, and buried himself two feet deep in clay.

p. 214, which lacks the tag: Lately, some DNA tags (a.k.a. DNA watermarks) have gotten rather elaborate, encoding names, e-mail addresses, or famous quotations—things nature couldn't have inserted by chance. One research team headed by Craig Venter encoded the following quotes in A's, C's, G's, and T's, then wove them into a synthetic genome that they created from scratch and inserted into a bacterium:

To live, to err, to fall, to triumph, to recreate life out of life.
> — James Joyce, *A Portrait of the Artist as a Young Man*

See things not as they are, but as they might be.
> — From *American Prometheus*, a book about Robert Oppenheimer

What I cannot build, I cannot understand.
> — Richard Feynman (the words written on his blackboard at the time of his death)

Unfortunately Venter bungled the last quote. Feynman actually wrote, "What I cannot create, I do not understand." Venter also ran into trouble with the Joyce quote. Joyce's family (which controls his estate) is reputedly stingy about letting anyone (including a bacterium) quote him without express written permission.

p. 218, millions of tons of vaporized rock per second: Compared to Mount Saint Helens, Toba spewed two thousand times more crap into the air. Of volcanoes worldwide, Toba is one of the few rivals to the giga-volcano currently smoldering beneath Wyoming, which will blow Yellowstone and everything around it sky-high someday.

Chapter 11: *Size Matters*

p. 227, accidental, unrelated to his genius: Stephen Jay Gould gives a highly entertaining rendition of the story of Cuvier's autopsy in *The Panda's Thumb*. Gould also wrote a masterly, two-part article about the life of Jean-Baptiste Lamarck—whom we'll meet in chapter 15—in his collection *The Lying Stones of Marrakech*.

p. 229, the cost is a punier brain: Partly to determine how and why hobbits shrank, scientists are currently drilling a hobbit tooth to extract DNA. It's a dicey procedure, since hobbits (unlike Neanderthals) lived in exactly the sort of tropical climate that degrades DNA the quickest. Attempts to extract hobbit DNA have always failed so far.

Studying hobbit DNA should help scientists determine if it really belongs in the genus *Homo*, a contentious point. Until 2010, scientists knew of only two other *Homo* species—Neanderthals, and possibly hobbits— still alive when *Homo sapiens* began overrunning the planet. But scientists recently had to add another to the list, the Denisovans (dun-EE-suh-vinz), named after a cave in cold Siberia where a five-year-old girl died tens of thousands of years ago. Her bones looked Neanderthal when scientists discovered them amid ancient layers of dirt and goat feces in 2010, but DNA extracted from a knucklebone shows enough distinctions to count as a separate line of *Homo*—the first extinct species discovered solely through genetic (not anatomical) evidence.

Traces of Denisovan DNA are found today in Melanesians, the people who originally settled the islands between New Guinea and Fiji. Apparently the Melanesians ran into Denisovans somewhere on the long haul from Africa to the South Seas and, as their ancestors had with Neanderthals, interbred with them. Today Melanesians have up to 8 percent non– *Homo sapiens* DNA. But beyond these clues, the Denisovans remain a mystery.

p. 233, scattered the ashes: Want more? Galileo's finger, Oliver Cromwell's skull, and Jeremy Bentham's entire decapitated head (including its freakishly shrunken skin) all went on posthumous display over the centuries. Thomas Hardy's heart reportedly got eaten by a cat. Phrenologists stole Joseph Haydn's head just before burial, and cemetery workers stole Franz Schubert's "larvae-laden" hair while transferring him to a new grave. Someone even held a jar in front of Thomas Edison's mouth during his death rattle to capture his last breath. The jar immediately went on exhibit in a museum.

I could probably go on for another page listing famous body parts that found new life—Percy Bysshe Shelley's heart, Grover Cleveland's cancerous jaw, supposed bits of Jesus's foreskin (the Divine Prepuce)—but let me wrap up by pointing out that there are no legs to the persistent rumor that the Smithsonian Institution owns John Dillinger's penis.

p. 235, pack the cortex with neurons: The overall genetic algorithm to add bulk and density to the brain might be amazingly simple. The biologist Harry Jerison has proposed the following example. Imagine a stem cell whose DNA programs it to "divide thirty-two times, then stop." If no cells die, you'd end up with 4,294,967,296 neurons. Now imagine tweaking that code to "divide thirty-four times, then stop." This would lead to two more doublings, or 17,179,869,184 neurons.

The difference between 4.3 billion neurons and 17.2 billion neurons, Jerison notes, would be roughly the difference between the chimpanzee cortex population and the human cortex population. "The code may seem overly simple," Jerison says, but "instructions that are significantly more complex may be beyond the capacity of genes to encode information."

p. 241, knew Mormon theology in lobotomizing detail: It's unclear if Peek, a devout Mormon, knew about the rift that genetic archaeology has recently opened within the Church of Latter-Day Saints. Mormons have traditionally believed—ever since Joseph Smith, just fourteen years old, copied down Jehovah's very words in 1820—that both Polynesians and American Indians descended from a doughty Jewish prophet, Lehi, who sailed from Jerusalem to America in 600 BC. Every DNA test ever conducted on these peoples disagrees on this point, however: they're not Middle Eastern in the slightest. And this contradiction not only invalidates the literalness of the Mormon holy books, it upsets the complicated Mormon eschatology about which brown peoples will be saved during end times, and which groups therefore need proselytizing to in the meantime. This finding has caused a lot of soul wringing among some Mormons, especially university scientists. For some, it crushed their faith. Most run-of-the-mill Latter-Day Saints presumably either don't know or have absorbed the contradiction and moved on.

p. 243, certainly doesn't capture him: For an account that does succeed in capturing Peek's talents, see "Inside the Mind of a Savant," by Donald Treffert and Daniel Christensen, in the December 2005 issue of *Scientific American*.

Chapter 12: *The Art of the Gene*

p. 247, awfully awkward method of replication: The warped-zipper model with its alternating left and right helixes actually debuted twice in 1976 (more simultaneous discovery). First a team in New Zealand published the idea. Shortly thereafter, a team working independently in India came forward with two warped-zipper models, one identical to the New Zealanders' and one with some of the A's, C's, G's, and T's flipped upside down. And true to the cliché of intellectual rebels everywhere, almost all the members of both teams were outsiders to molecular biology and had no preconceived notions that DNA had to be a double helix. One New Zealander wasn't a professional scientist, and one Indian contributor had never heard of DNA!

p. 250, musical biases as ruthless as any hipster's: Monkeys either ignore human music or find it irritating, but recent studies with cotton-top tamarins, monkeys in South America, have confirmed that they respond strongly to music tailored for them. David Teie, a cellist based in Maryland, worked with primatologists to compose music based on the calls that tamarins use to convey fear or contentment. Specifically, Teie patterned his work on the rising and falling tones in the calls, as well as their duration, and when he played the various opuses, the cotton-tops showed visible signs of relaxation or anxiety. Showing a good sense of humor, Teie commented to a newspaper, "I may be just a schmo to you. But, man, to monkeys I am Elvis."

p. 255, the three times he owned up to crying: Because you're dying to know, the first time Rossini cried was when his first opera flopped. The bawling with Paganini was the second. The third and final time, Rossini, a bona fide glutton, was boating with friends when—horror of horrors—his picnic lunch, a delectable truffled turkey, fell overboard.

p. 260, the church finally forgave him and permitted burial: Biographies of Paganini are surprisingly scarce in English. One short and lively introduction to his life—with lots of details about his illnesses and postmortem travails—is *Paganini*, by John Sugden.

p. 262, throbbing-red baboon derrieres: For whatever reason, some classic American writers paid close attention to debates in the early 1900s about sexual selection and its role in human society. F. Scott Fitzgerald, Ernest Hemingway, Gertrude Stein, and Sherwood Anderson all addressed the animal aspects of courtship, male passion and jealousy, sexual ornamentation, and so on. In a similar way, genetics itself shook some of these writers. In his fascinating *Evolution and the "Sex Problem,"* Bert Bender writes that "although Mendelian genetics was a welcome discovery for Jack London, who heartily embraced it as a rancher who practiced selective breeding, others, such as Anderson, Stein, and Fitzgerald, were deeply disturbed." Fitzgerald especially seemed obsessed with evolution, eugenics, and heredity. Bender points out that he makes constant references to eggs in his work (West Egg and East Egg are but two examples), and Fitzgerald once wrote about "the cut-in system at dances, which favors the survival of the fittest." Even Gatsby's "old sport," his nickname for Nick Carraway, the narrator of *The Great Gatsby*, probably has its roots in the early habit among geneticists of calling mutants "sports."

p. 263, bone deformities and seizure disorders: Armand Leroi's *Mutants* explores in more detail what specific disease Toulouse-Lautrec might have had, and also the effect on his art. In fact I highly recommend the book overall for its many fascinating tales, like the anecdote about the lobster claw–like birth defects mentioned in chapter 1.

p. 264, looking rather apish: The lip was more obvious in pictures of men, but women didn't escape these genes. Reportedly, Marie Antoinette, part of another branch of the family, had strong traces of the Hapsburg lip.

Chapter 13: *The Past Is Prologue — Sometimes*

p. 282, Lincoln, fifty-six when assassinated: Funnily enough, Lincoln's assassin got caught up in a genetic contretemps of his own in the 1990s. Two historians at the time were peddling a theory that it wasn't John Wilkes Booth but an innocent bystander whom Union soldiers tracked, captured, and killed in Bowling Green, Virginia, in 1865, twelve days after the assassination. Instead, the duo argued, Booth gave the troops the slip, fled west, and lived thirty-eight increasingly wretched years in Enid, Oklahoma, before committing suicide in 1903. The only way to know was to exhume the body in Booth's grave, extract DNA, and test it against his living relatives. The caretakers of the cemetery refused, however, so Booth's family (spurred on by the historians) sued. A judge denied their petition, partly because the technology at the time probably couldn't have resolved the matter; but the case could in theory be reopened now.

For more details on Booth's and Lincoln's DNA see, well, *Abraham Lincoln's DNA and Other Adventures in Genetics*, by Philip R. Reilly, who sat on the original committee that studied the feasibility of testing Lincoln.

p. 285, Jewish tradition bungled this story: Overall, though, Jewish people were acute observers of hereditary phenomena. By AD 200, the Talmud included an exemption to not circumcise a young boy if two older brothers had bled to death after their circumcisions. What's more, Jewish law later exempted half brothers of the deceased as well — but only if they had the same mother. If the half brother shared a father, then the circumcision should proceed. The children of women whose sisters' babies had bled to death also received exemptions, but not the children of men whose brothers' babies had. Clearly Jewish people understood a long time ago that the disease in question — probably hemophilia, an inability of blood to clot — is a sex-linked disease that affects mostly males but gets passed down through the mother.

p. 291, severe lactose intolerance: If the world were just, we would call the condition not "lactose intolerance" but "lactose tolerance," since the ability to digest milk is the oddity and only came about because of a recent mutation. Actually, two recent mutations, one in Europe, one in Africa. In both cases the mutation disabled a region on chromosome two that, in adults, should halt production of the enzyme that digests lactose, a sugar in milk. And while the European one came first historically (about 7,000 BC), one scientist said the African gene spread especially quickly: "It is basically the strongest signal of selection ever observed in any genome, in any study, in any population in the world." Lactose tolerance is also a wonderful example of gene-culture coevolution, since the ability to digest milk would not have benefited anyone before the domestication of cattle and other beasts, which gave a steady supply of milk.

Chapter 14: *Three Billion Little Pieces*

p. 295, tally the sequence: If you like to get your fingernails dirty, you can visit http://samkean.com/thumb-notes for the gritty details of Sanger's work.

p. 299, giving Venter an F: Not a biology teacher but an English teacher inspired this devotion from Venter. And what an English teacher! It was Gordon Lish, later famous as the editor of Raymond Carver.

p. 307, Einstein's brain, to see if someone could sequence its DNA after all: Celera people had a thing about celebrity DNA kitsch. According to James Shreeve's riveting book, *The Genome War*, the lead architect of Celera's ingenious supercomputer program kept on his office bookshelf a "pus-stained Band-Aid" in a test tube—an homage to Friedrich Miescher. By the way, if you're interested in a long insider's account of the HGP, Shreeve's book is the best written and most entertaining I know of.

p. 309, in early 2003: Actually this declaration of being "done" was arbitrary as well. Work on some parts of the human genome—like the hypervariable MHC region—continued for years, and cleanup work continues even today as scientists tidy up small mistakes and sequence segments that, for technical reasons, can't be sequenced by conventional means. (For instance, scientists usually use bacteria for the photocopying step. But some stretches of human DNA happen to poison bacteria, so bacteria delete them instead of copying, and they disappear.) Finally, scientists have not yet tackled telomeres and centromeres, segments that, respectively, make up the ends and central girdles of chromosomes, because those regions are too repetitive for conventional sequencers to make sense of.

So why did scientists declare the work done in 2003? The sequence did meet at that point a sensible definition of done: fewer than one error per 10,000 bases over 95 percent of the DNA regions that contain genes. Just as important, though, PR-wise, early 2003 was the fiftieth anniversary of Watson and Crick's discovery of the double helix.

p. 310, Venter can claim he won the genome war after all: On the other hand, it might actually *bolster* Venter's reputation in a topsy-turvy way if he lost out on the Nobel. The loss would confirm his status as an outsider (which endeared him to many people) and would give historians something to argue about for generations, making Venter the central (and perhaps tragic) figure in the HGP story.

Watson's name doesn't come up often in discussions about the Nobel, but he arguably deserves one for compelling Congress—not to mention the majority of geneticists in the country—to give sequencing a chance. That said, Watson's recent gaffes, especially his disparaging comments about the intelligence of Africans (more on this later), may have killed his chances. It sounds crude for me to say, but the Nobel committee might wait for Watson to croak before giving out any HGP-related prizes.

If either Watson or Sulston won, it would be his second Nobel, matching Sanger as the only double winner in medicine/physiology. (Sulston won his prize for worm work in 2002.) Like Watson, though, Sulston has entangled himself in some controversial politics. When WikiLeaks founder Julian Assange got arrested in 2010—charged with sexual assault in Sweden, the Nobel homeland—Sulston offered many thousands of pounds for bail. It seems Sulston's commitment to the free and unimpeded flow of information doesn't stop at the laboratory door.

p. 313, to conceal these results didn't succeed: An amateur scientist named Mike Cariaso outed Watson's *apoE* status by taking advantage of genetic hitchhiking. Again, because of hitchhiking, each different version of a gene will have, purely by chance, certain versions of other genes associated with it—genes that travel with it down the generations. (Or if there aren't genes nearby, each version of the gene will at least have certain junk DNA associated with it.) So if you wanted to know which version of *apoE* someone had, you could look at *apoE* itself or, just as good, look at the genes flanking it. The scientists in charge of scrubbing this information from Watson's genome of course knew this, and they erased the information near *apoE*. But they didn't erase enough. Cariaso realized this mistake and, simply by looking up Watson's publicly available DNA, figured out his *apoE* status.

As described in *Here Is a Human Being*, by Misha Angrist, Cariaso's revelation was a shock, not least because he was something of an expat bum: "The facts were inescapable: The Nobel Prize–winning [Watson] had asked that of his 20,000+ genes...just one lousy gene—one!—not be made public. This task was left to the molecular brain trust at Baylor University, one of the top genome centers in the world.... But the Baylor team was outfoxed by a thirty-year-old autodidact with a bachelor's degree who preferred to spend most of his time on the Thai-Burmese border distributing laptops and teaching kids how to program and perform searches in Google."

Chapter 15: *Easy Come, Easy Go?*

p. 319, banish from their field forever: History has a funny sense of humor. Darwin's grandfather Erasmus Darwin, a physician, actually published an independent and fairly outlandish theory of evolution (in verse, no less) that resembled Lamarck's. Samuel Taylor Coleridge even coined the word "Darwining" to pooh-pooh such speculation. Erasmus also started the family tradition of raising the hackles of religious folk, as his work appeared on the papal index of banned books.

In another irony, just after Cuvier died, he himself got smeared in the same way he'd smeared Lamarck. Based on his views, Cuvier became indelibly linked to catastrophism and an antievolutionary outlook on natural history. So when Charles Darwin's generation needed a foil to represent stodgy old thinking, the pumpkin-headed Frenchman was perfect, and Cuvier's reputation suffers even today for their lancing. Payback's a bitch.

p. 331, the tempestuous ex-wife of (among others) composer Gustav Mahler: Alma Mahler also had the good taste to marry painter Gustav Klimt and Bauhaus designer Walter Gropius, among others. She became such a notorious jezebel in Vienna that Tom Lehrer wrote a song about her. The refrain goes: "Alma, tell us! / All modern women are jealous. / Which of your magical wands / Got you Gustav and Walter and Franz?" I have a link to the full lyrics on my website.

Chapter 16: *Life as We Do (and Don't) Know It*

p. 341, before she was ever born: Since Dolly, scientists have cloned cats, dogs, water buffalo, camels, horses, and rats, among other mammals. In 2007 scientists created embryonic clones from adult monkey cells and let them develop enough to see different tissues. But the scientists snuffed the embryos before they came to term, so it's unclear whether the monkey clones would have developed normally. Primates are harder to clone than other species because removing the nucleus of the donor egg (to make way for the clone's chromosomes) tears out some of the special apparatus that primate cells need to divide properly. Other species have more copies of this apparatus, called spindle fibers, than primates do. This remains a major technical obstacle to human cloning.

p. 346, letting politics color their science: Psychoanalyze this how you will, but both James Watson and Francis Crick have stuck their size tens into their mouths with impolitic public comments about race, DNA, and intelligence. Crick lent support to research in the 1970s on why some racial groups had—or, more truthfully, were tested to have—higher or lower IQs than others. Crick thought we could form better social policies if we knew that certain groups had lower intellectual ceilings. He also said, more bluntly, "I think it likely that more than half the difference between the average IQ of American whites and Negroes is due to genetic reasons."

Watson's gaffe came in 2007 while on tour to promote his charmingly titled autobiography, *Avoid Stupid People*. At one point he proclaimed, "I'm inherently gloomy about the prospect of Africa," since "social policies are based on the fact that their intelligence is the same as ours. Whereas all the testing says, not really." After the media flagellated him, Watson lost his job (as head of Cold Spring Harbor Laboratory, Barbara McClintock's old lab) and more or less retired in semidisgrace.

It's hard to know how seriously to take Watson here, given his history of saying crude and inflammatory things—about skin color and sex drive, about women ("People say it would be terrible if we made all girls pretty. I think it would be great"), about abortion and sexuality ("If you could find the gene which determines sexuality and a woman decides she doesn't want a homosexual child, well, let her"), about obese people ("Whenever you interview fat people, you feel bad, because you know you're not going to hire them"), et cetera. The Harvard black studies scholar Henry Louis Gates Jr. later probed Watson about the Africa comments in a private meeting and concluded that Watson wasn't so much racist as "racialist"— someone who sees the world in racial terms and believes that genetic gaps may exist between racial groups. Gates also noted, though, that Watson believes that if such gaps do exist, the *group* differences shouldn't bias our views toward talented *individuals*. (It's analogous to saying that black people may be better at basketball, but the occasional Larry Bird can still thrive.) You can read Gates's thoughts at http://www.washingtonpost.com /wp-dyn/content/article/2008/07/10/AR2008071002265.html.

As always, DNA had the last word. Along with his autobiography, Watson released his genome to the public in 2007, and some scientists decided to plumb it for markers of ethnicity. Lo and behold, they discovered that Watson might have, depending on the accuracy of his sequence, up to sixteen times more genes from black Africans than a typical Caucasian has—the genetic equivalent of a black great-grandfather.

p. 352, an extra pair of chromosomes and inserting them into embryos: Among other people, Nicholas Wade makes this suggestion in *Before the Dawn*, a masterly tour of all aspects of human origins— linguistic, genetic, cultural, and otherwise.

Selected Bibliography

Here's a list of books and papers I consulted while writing *The Violinist's Thumb*. Anything marked with an asterisk I recommend especially. I've annotated the ones I recommend especially for further reading.

Chapter 1: *Genes, Freaks, DNA*

Bondeson, Jan. *A Cabinet of Medical Curiosities*. W. W. Norton, 1999.
*Contains an astounding chapter on maternal impressions, including the fish boy of Naples.

Darwin, Charles. *On the Origin of Species*. Introduction by John Wyon Burrow. Penguin, 1985.

———. *The Variation of Animals and Plants Under Domestication*. J. Murray, 1905.

Henig, Robin Marantz. *The Monk in the Garden*. Houghton Mifflin Harcourt, 2001.
*A wonderful general biography of Mendel.

Lagerkvist, Ulf. *DNA Pioneers and Their Legacy*. Yale University Press, 1999.

Leroi, Armand Marie. *Mutants: On genetic variety and the human body*. Penguin, 2005.
*A fascinating account of maternal impressions, including the lobster claw–like birth defects.

Chapter 2: *The Near Death of Darwin*

Carlson, Elof Axel. *Mendel's Legacy*. Cold Spring Harbor Laboratory Press, 2004.
*Loads of anecdotes about Morgan, Muller, and many other key players in early genetics, by a student of Muller's.

Endersby, Jim. *A Guinea Pig's History of Biology*. Harvard University Press, 2007.

*A marvelous history of the fly room. One of my favorite books ever, in fact. Endersby also touches on Darwin's adventures with gemmules, Barbara McClintock, and other tales.

Gregory, Frederick. *The Darwinian Revolution*. DVDs. Teaching Company, 2008.

Hunter, Graeme K. *Vital Forces*. Academic Press, 2000.

Kohler, Robert E. *Lords of the Fly*. University of Chicago Press, 1994.
*Includes details about Bridges's private life, like the anecdote about his Indian "princess."

Steer, Mark, et al., eds. *Defining Moments in Science*. Cassell Illustrated, 2008.

Chapter 3: *Them's the DNA Breaks*

Hall, Eric J., and Amato J. Giaccia. *Radiobiology for the Radiologist*. Lippincott Williams and Wilkins, 2006.
*A detailed but readable account of how exactly radioactive particles batter DNA.

Hayes, Brian. "The Invention of the Genetic Code." *American Scientist*, January–February 1998.
*An entertaining rundown of early attempts to decipher the genetic code.

Judson, Horace F. *The Eighth Day of Creation*. Cold Spring Harbor Laboratory Press, 2004.
*Includes the story of Crick not knowing what *dogma* meant.

Seachrist Chiu, Lisa. *When a Gene Makes You Smell Like a Fish*. Oxford University Press, 2007.

Trumbull, Robert. *Nine Who Survived Hiroshima and Nagasaki*. Dutton, 1957.
*For a fuller account of Yamaguchi's story—and for eight other equally riveting tales—I can't recommend this book enough.

Chapter 4: *The Musical Score of DNA*

Flapan, Erica. *When Topology Meets Chemistry*. Cambridge University Press, 2000.

Frank-Kamenetskii, Maxim D. *Unraveling DNA*. Basic Books, 1997.

Gleick, James. *The Information*. HarperCollins, 2011.

Grafen, Alan, and Mark Ridley, eds. *Richard Dawkins*. Oxford University Press, 2007.

Zipf, George K. *Human Behavior and the Principle of Least Effort*. Addison-Wesley, 1949.

———. *The Psycho-biology of Language*. Routledge, 1999.

Chapter 5: *DNA Vindication*

Comfort, Nathaniel C. "The Real Point Is Control." *Journal of the History of Biology* 32 (1999): 133–62.

*Comfort is the scholar most responsible for challenging the mythic, fairy-tale version of Barbara McClintock's life and work.

Truji, Jan. *The Soul of DNA*. Llumina Press, 2004.
*For a more detailed account of Sister Miriam, I highly recommend this book, which chronicles her life from its earliest days to the very end.

Watson, James. *The Double Helix*. Penguin, 1969.
*Watson recalls multiple times his frustration over the different shapes of each DNA base.

Chapter 6: *The Survivors, the Livers*

Hacquebord, Louwrens. "In Search of *Het Behouden Huys*." *Arctic* 48 (September 1995): 248–56.

Veer, Gerrit de. *The Three Voyages of William Barents to the Arctic Regions*. N.p., 1596.

Chapter 7: *The Machiavelli Microbe*

Berton, Pierre. *Cats I Have Known and Loved*. Doubleday Canada, 2002.

Dulbecco, Renato. "Francis Peyton Rous." In *Biographical Memoirs*, vol. 48. National Academies Press, 1976.

McCarty, Maclyn. *The Transforming Principle*. W. W. Norton, 1986.

Richardson, Bill. *Scorned and Beloved: Dead of Winter Meetings with Canadian Eccentrics*. Knopf Canada, 1997.

Villarreal, Luis. "Can Viruses Make Us Human?" *Proceedings of the American Philosophical Society* 148 (September 2004): 296–323.

Chapter 8: *Love and Atavisms*

Bondeson, Jan. *A Cabinet of Medical Curiosities*. W. W. Norton, 1999.
*A marvelous section on human tails, from a book chock-full of gruesome tales from the history of anatomy.

Isoda, T., A. Ford, et al. "Immunologically Silent Cancer Clone Transmission from Mother to Offspring." *Proceedings of the National Academy of Sciences of the United States of America* 106, no. 42 (October 20, 2009): 17882–85.

Villarreal, Luis P. *Viruses and the Evolution of Life*. ASM Press, 2005.

Chapter 9: *Humanzees and Other Near Misses*

Rossiianov, Kirill. "Beyond Species." *Science in Context* 15, no. 2 (2002): 277–316.
*For more on Ivanov's life, this is the most authoritative and least sensationalistic source.

Chapter 10: *Scarlet A's, C's, G's, and T's*

Barber, Lynn. *The Heyday of Natural History*. Cape, 1980.
  *A great source for information about the Bucklands, *père* and *fils*.
Carroll, Sean B. *Remarkable Creatures*. Houghton Mifflin Harcourt, 2009.
Finch, Caleb. *The Biology of Human Longevity*. Academic Press, 2007.
Finch, Caleb, and Craig Stanford. "Meat-Adaptive Genes Involving Lipid Metabolism Influenced Human Evolution." *Quarterly Review of Biology* 79, no. 1 (March 2004): 3–50.
Sommer, Marianne. *Bones and Ochre*. Harvard University Press, 2008.
Wade, Nicholas. *Before the Dawn*. Penguin, 2006.
  *A masterly tour of all aspects of human origins.

Chapter 11: *Size Matters*

Gould, Stephen Jay. "Wide Hats and Narrow Minds." In *The Panda's Thumb*. W. W. Norton, 1980.
  *A highly entertaining rendition of the story of Cuvier's autopsy.
Isaacson, Walter. *Einstein: His Life and Universe*. Simon and Schuster, 2007.
Jerison, Harry. "On Theory in Comparative Psychology." In *The Evolution of Intelligence*. Psychology Press, 2001.
Treffert, D., and D. Christensen. "Inside the Mind of a Savant." *Scientific American*, December 2005.
  *A lovely account of Peek by the two scientists who knew him best.

Chapter 12: *The Art of the Gene*

Leroi, Armand Marie. *Mutants: On Genetic Variety and the Human Body*. Penguin, 2005.
  *This marvelous book discusses in more detail what specific disease Toulouse-Lautrec might have had, and also the effect on his art.
Sugden, John. *Paganini*. Omnibus Press, 1986.
  *One of the few biographies of Paganini in English. Short, but well done.

Chapter 13: *The Past Is Prologue — Sometimes*

Reilly, Philip R. *Abraham Lincoln's DNA and Other Adventures in Genetics*. Cold Spring Harbor Laboratory Press, 2000.
  *Reilly sat on the original committee that studied the feasibility of testing Lincoln's DNA. He also delves into the testing of Jewish people's DNA, among other great sections.

Chapter 14: *Three Billion Little Pieces*

Angrist, Misha. *Here Is a Human Being*. HarperCollins, 2010.
  *A lovely and personal rumination on the forthcoming age of genetics.
Shreeve, James. *The Genome War*. Ballantine Books, 2004.

*If you're interested in an insider's account of the Human Genome Project, Shreeve's book is the best written and most entertaining I know of.

Sulston, John, and Georgina Ferry. *The Common Thread*. Joseph Henry Press, 2002.

Venter, J. Craig. *A Life Decoded: My Genome—My Life*. Penguin, 2008.
*The story of Venter's entire life, from Vietnam to the HGP and beyond.

Chapter 15: *Easy Come, Easy Go?*

Gliboff, Sander. "Did Paul Kammerer Discover Epigenetic Inheritance? No and Why Not." *Journal of Experimental Zoology* 314 (December 15, 2010): 616–24.

Gould, Stephen Jay. "A Division of Worms." *Natural History*, February 1999.
*A masterly two-part article about the life of Jean-Baptiste Lamarck.

Koestler, Arthur. *The Case of the Midwife Toad*. Random House, 1972.

Serafini, Anthony. *The Epic History of Biology*. Basic Books, 2002.

Vargas, Alexander O. "Did Paul Kammerer Discover Epigenetic Inheritance?" *Journal of Experimental Zoology* 312 (November 15, 2009): 667–78.

Chapter 16: *Life as We Do (and Don't) Know It*

Caplan, Arthur. "What If Anything Is Wrong with Cloning a Human Being?" *Case Western Reserve Journal of International Law* 35 (Fall 2003): 69–84.

Segerstråle, Ullica. *Defenders of the Truth*. Oxford University Press, 2001.

Wade, Nicholas. *Before the Dawn*. Penguin, 2006.
*Among other people, Nicholas Wade suggested adding the extra pair of chromosomes.

INDEX

Italic page numbers refer to illustrations.

Sam Kean spent years collecting mercury from broken thermometers as a child and now he is a writer in Washington DC. His work has appeared in the *New York Times Magazine*, *Slate*, and *New Scientist*. In 2009 he was a runner-up for the National Association of Science Writers' Award for best science writer under the age of thirty. His first book, *The Disappearing Spoon*, was a *New York Times* bestseller and was shortlisted for the Royal Society's Winton Prize for Science Books 2011.